Beyond the Messier
and
Caldwell Catalogues

by

Martin P Nicholson

Introduction

If you subscribe to one of the magazines that cater for amateur astronomers you will soon realise that the same few objects are mentioned again and again as suitable targets for owners of small telescopes. I am, of course, referring to those listed in the Messier catalogue. However there is a limit to how many times an object can be viewed or imaged before boredom sets in. It is hardly surprising that more active observers started looking for advice on possible deep-sky objects "beyond the Messier catalogue".

Probably the best known of these new lists was the Caldwell catalogue compiled by the late Sir Patrick Caldwell-Moore. This is a list of 109 galaxies, nebulae and star clusters scattered across the entire sky that were not listed by Messier. As a source of objects of interest the Caldwell catalogue was unobjectionable but as with so many things in science the "devil was in the detail".

Sir Patrick decided, for reasons that were never entirely clear, to give Caldwell numbers to objects that already had well-known and long-standing designations attached to them. This caused significant disquiet throughout both the amateur and professional astronomical communities. A surprising number of newcomers to the hobby assumed (entirely incorrectly) that Sir Patrick had discovered these 109 objects and was therefore entitled to claim naming rights to them! Another problem was the number of mistakes in the data tables associated with the Caldwell catalogue. I can still remember looking with disbelief at both the quoted size and magnitude of some of the objects.

Some people – more in sorrow that anger – felt that Sir Patrick was guilty of serious self-aggrandisement in attaching his name to objects other people had discovered. One thing is certain: Caldwell numbers are seldom used by experienced astronomers.

Many other lists of astronomical targets have been compiled. Perhaps the best known of these is the Herschel 400 (https://www.astroleague.org/al/obsclubs/herschel/hers400.html). All 400 targets can be seen with a 6 inch telescope from a moderately dark site. These lists seem to have one important feature in common – they are sub-sets extracted from much larger catalogues using a range of selection criteria that are not always well explained.

> **My lists of targets are not like these others. I have discovered all these objects myself: either by direct observation or via data mining.**

For over 20 years I had always assumed that all the bright, close double stars had long since been discovered. I knew that the details of all the catalogued binary and double stars could be found in the "The Washington Visual Double Star Catalogue" (WDS) and that the WDS could be accessed at this web address:

http://vizier.hia.nrc.ca/viz-bin/VizieR-3?-source=B/wds/wds

The catalogue confirmed that while new systems were still being discovered, it was in identifying common proper motion pairs where most of the research activity was concentrated. The lack of newly discovered bright double stars meant that authors were almost bound both to identify the same few dozen double stars as "celestial highlights" and to use essentially identical third-party descriptions for the more routine systems in their check-lists of possible amateur targets. The resultant high degree of overlap between the different books and web sites written about double stars has been a source of mild frustration to me for some years.

In April 2014 the Journal of Double Star Observations contained a brief article by Bryant in which he announced the discovery of a new double star in the constellation of Perseus.

http://www.jdso.org/volume10/number2/Bryant_105_106.pdf

TVB 1 was said to have a separation of 14.1 arc seconds at a position angle of 22 degrees.

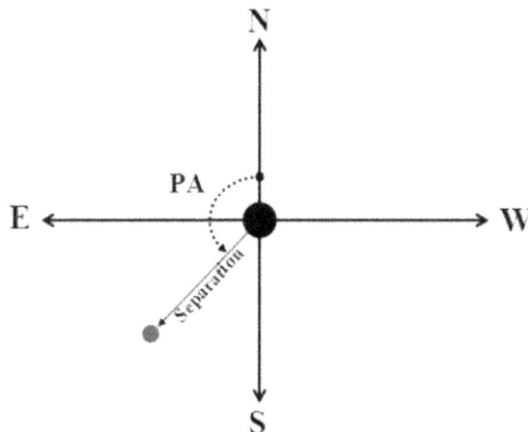

Fig. 1 - The key features of a double star

What surprised me most about this discovery was that this bright (magnitudes 10.3 and 10.6) double star had not been identified and formally catalogued many years ago. A secondary magnitude of 10.6 puts it comfortably into the top 20% of double stars by brightness and the separation is fairly typical of the new catalogue entries previously credited to the great double star astronomers of the 19th and 20th centuries.

I had already developed some fairly sophisticated astronomical data mining techniques so it only required some minor modification to the software I had written some years previously to conduct a systematic search for bright new double stars that were not listed in the standard catalogues.

Until quite recently I would always have made these discoveries available via informal posts to on-line newsgroups or by arranging for them to be published in a relevant society journal. Unfortunately in recent years there has been an increasing tendency for unscrupulous astronomers, both amateur and professional, to claim that neither method constitutes "publication" in the scientifically accepted sense of the word. They would then publish essentially identical results in peer reviewed magazines and claim – and be granted – both discovery credit and naming rights by the custodians of the standard catalogues.

I was left with no choice other than to publish my results in a peer reviewed book. I owe a significant debt to the anonymous peer-reviewer whose helpful suggestion significantly improved this study.

Nicholson #1 to #120

These 120 double stars fall into three distinct groups –

- Bright and close double stars (40 examples)
- Celestial twins
- Colour-contrasting double stars

BRIGHT AND CLOSE DOUBLE STARS	NICHOLSON #			
40 examples	10	27	68	92
	12	28	74	93
	19	31	75	97
	20	32	76	99
	21	38	77	102
	22	41	78	110
	23	43	80	111
	24	52	81	112
	25	54	85	118
	26	56	89	119

CELESTIAL TWINS	NICHOLSON #				
41 examples	3	34	57	83	120
	5	35	58	87	
	6	37	59	88	
	7	42	60	90	
	8	45	61	94	
	13	46	63	95	
	14	47	64	108	
	15	48	67	109	
	18	51	71	114	
	29	53	82	115	

COLOUR CONTRASTING DOUBLE STARS	NICHOLSON #			
39 examples	1	39	70	101
	2	40	72	103
	4	44	73	104
	9	49	79	105
	11	50	84	106
	16	55	86	107
	17	62	91	113
	30	65	96	116
	33	66	98	117
	36	69	100	

Fig. 2 - The three sub-types from Nicholson #1 to #120

Explanation of the data tables:

1. # – The Nicholson number of the newly discovered double star.
2. H, M and S – The right ascension of the primary star in hours, minutes and seconds.
3. D, M and S – The declination of the primary star in degrees, minutes and seconds.
4. MAG1 and MAG2 – The magnitudes of the primary and secondary stars.
5. SEP – The distance between the stars in arc seconds.
6. PA – The position angle between the stars in degrees.
7. B-V – The colour of the primary star (the magnitude in the B band minus the magnitude in the V band).
8. B-V – The colour of the secondary star (the magnitude in the B band minus the magnitude in the V band).
9. CON – The abbreviated name of the constellation in which the double star can be found.

And	Andromeda	Cyg	Cygnus	Pav	Pavo
Ant	Antlia	Del	Delphinus	Peg	Pegasus
Aps	Apus	Dor	Dorado	Per	Perseus
Aqr	Aquarius	Dra	Draco	Phe	Phoenix
Aql	Aquila	Equ	Equuleus	Pic	Pictor
Ara	Ara	Eri	Eridanus	Psc	Pisces
Ari	Aries	For	Fornax	PsA	Pisces Austrinus
Aur	Auriga	Gem	Gemini	Pup	Puppis
Boo	Bootes	Gru	Grus	Pyx	Pyxis
Cae	Caelum	Her	Hercules	Ret	Reticulum
Cam	Camelopardalis	Hor	Horologium	Sge	Sagitta
Cnc	Cancer	Hya	Hydra	Sgr	Sagittarius
CVn	Canes Venatici	Hyi	Hydrus	Sco	Scorpius
CMa	Canis Major	Ind	Indus	Scl	Sculptor
CMi	Canis Minor	Lac	Lacerta	Sct	Scutum
Cap	Capricornus	Leo	Leo	Ser	Serpens
Car	Carina	LMi	Leo Minor	Sex	Sextans
Cas	Cassiopeia	Lep	Lepus	Tau	Taurus
Cen	Centaurus	Lib	Libra	Tel	Telescopium
Cep	Cepheus	Lup	Lupus	Tri	Triangulum
Cet	Cetus	Lyn	Lynx	TrA	Triangulum Australe
Cha	Chamaleon	Lyr	Lyra	Tuc	Tucana
Cir	Circinus	Men	Mensa	UMa	Ursa Major
Col	Columba	Mic	Microscopium	UMi	Ursa Minor
Com	Coma Berenices	Mon	Monoceros	Vel	Vela
CrA	Corona Australis	Mus	Musca	Vir	Virgo
CrB	Corona Borealis	Nor	Norma	Vol	Volans
Crv	Corvus	Oct	Octans	Vul	Vulpecula
Crt	Crater	Oph	Ophiucus		
Cru	Crux	Ori	Orion		

<u>Fig. 3 - The three letter abbreviations plus the associated full constellation names</u>

The finder charts have dimensions of 300 by 300 arc seconds.

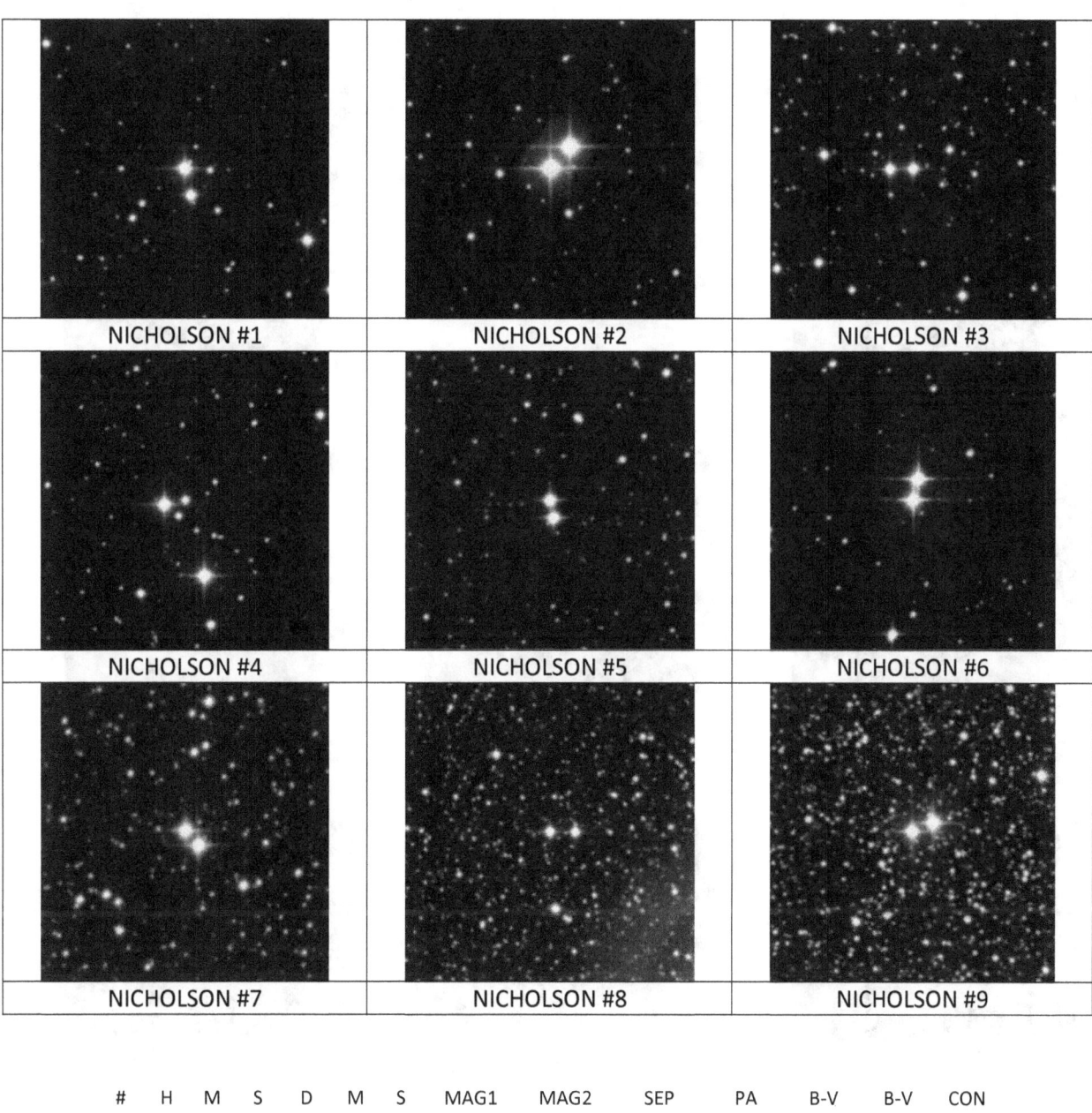

NICHOLSON #1   NICHOLSON #2   NICHOLSON #3

NICHOLSON #4   NICHOLSON #5   NICHOLSON #6

NICHOLSON #7   NICHOLSON #8   NICHOLSON #9

#	H	M	S	D	M	S	MAG1	MAG2	SEP	PA	B-V	B-V	CON
1	0	13	46	44	32	46	10.48	11.20	27.64	195.43	2.04	0.59	And
2	1	17	30	44	3	32	8.41	9.42	29.44	316.67	0.16	1.53	And
3	23	25	27	50	40	20	10.70	10.73	23.95	89.69	0.51	0.52	And
4	23	30	59	43	15	55	9.65	9.90	22.40	103.84	1.79	0.60	And
5	10	7	8	-38	48	1	10.92	10.94	17.96	190.17	0.48	0.44	Ant
6	10	32	7	-32	10	33	9.52	9.55	21.73	345.82	1.03	1.01	Ant
7	15	29	58	-73	23	38	10.30	10.32	20.13	221.88	1.30	1.28	Aps
8	18	58	29	17	24	31	11.02	11.30	25.85	269.08	0.25	0.24	Aql
9	19	40	7	11	17	2	9.11	9.44	23.56	291.48	0.05	1.33	Aql

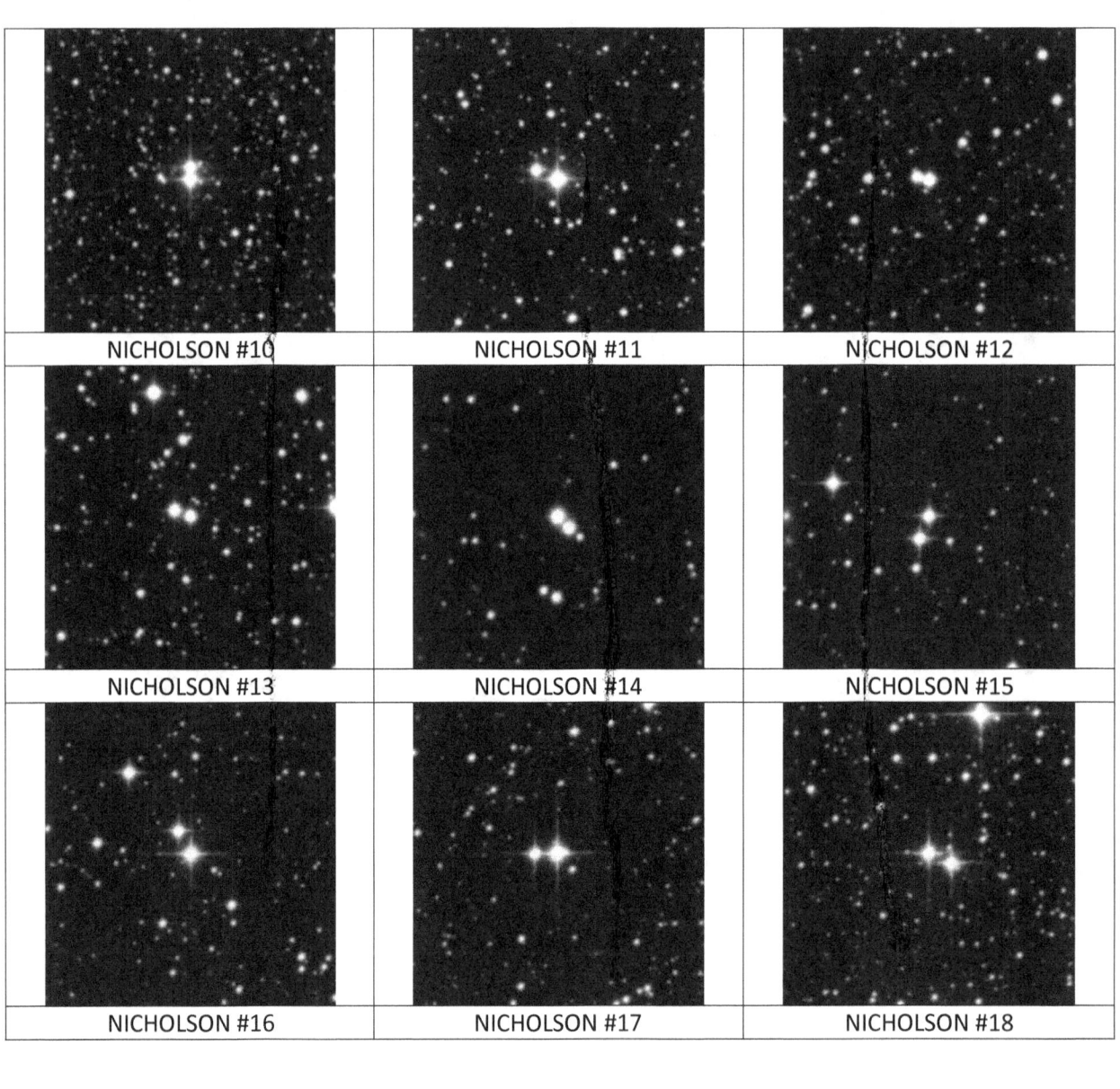

#	H	M	S	D	M	S	MAG1	MAG2	SEP	PA	B-V	B-V	CON
10	20	3	22	15	44	40	9.47	9.95	12.34	0.08	1.14	0.77	Aql
11	20	20	44	3	1	17	10.02	11.17	23.98	67.41	1.73	0.66	Aql
12	16	55	57	-57	15	13	10.17	10.89	12.40	79.13	0.77	0.98	Ara
13	5	11	59	43	20	6	10.84	10.99	17.40	70.09	0.43	0.43	Aur
14	4	57	10	67	30	21	10.73	11.10	16.04	222.52	0.37	0.37	Cam
15	6	33	53	-30	15	31	10.71	10.83	24.36	156.93	0.52	0.51	CMa
16	6	37	9	-17	46	32	10.07	10.90	26.08	30.22	1.24	0.24	CMa
17	7	15	45	-19	4	60	10.15	11.02	23.48	91.58	1.38	0.16	CMa
18	7	24	2	-24	21	20	10.01	10.07	25.26	246.09	0.91	0.90	CMa

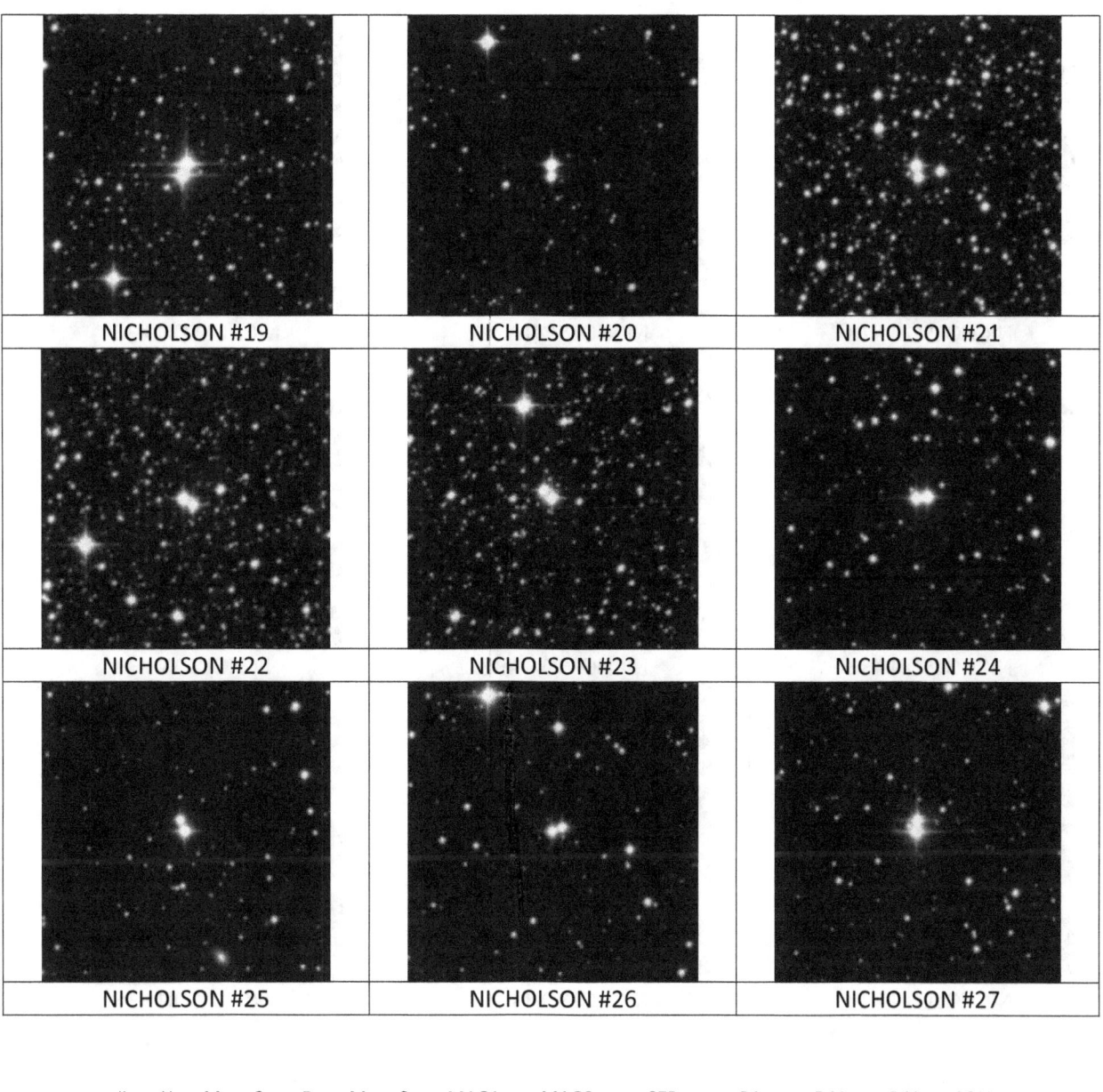

#	H	M	S	D	M	S	MAG1	MAG2	SEP	PA	B-V	B-V	CON
19	7	25	39	-26	16	31	9.37	9.61	11.87	158.61	0.42	0.41	CMa
20	9	24	51	-64	39	32	10.07	11.01	11.71	180.24	1.13	1.13	Car
21	10	6	53	-59	46	54	10.61	11.18	11.87	192.84	0.97	0.42	Car
22	10	34	58	-64	12	38	10.08	10.69	12.19	231.88	0.53	1.20	Car
23	10	54	1	-68	35	54	10.61	10.78	11.45	52.52	0.84	0.64	Car
24	0	41	17	53	16	1	10.50	10.51	11.48	278.66	0.47	0.51	Cas
25	0	48	32	49	56	36	10.40	10.91	12.17	21.57	0.93	0.52	Cas
26	0	55	27	48	20	1	10.70	10.82	11.25	289.37	0.63	0.68	Cas
27	1	4	38	51	23	52	10.00	10.83	12.28	344.78	1.21	0.37	Cas

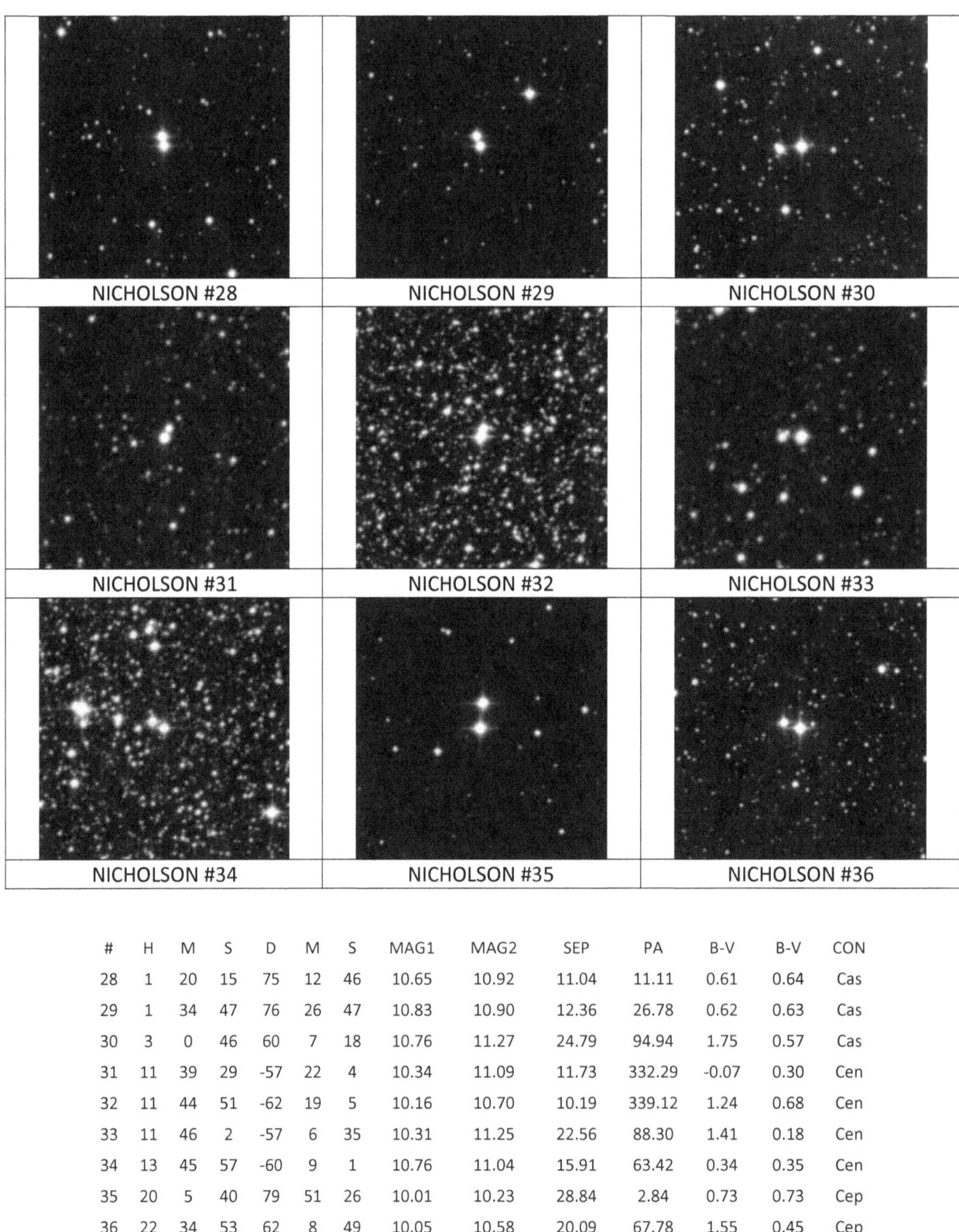

#	H	M	S	D	M	S	MAG1	MAG2	SEP	PA	B-V	B-V	CON
28	1	20	15	75	12	46	10.65	10.92	11.04	11.11	0.61	0.64	Cas
29	1	34	47	76	26	47	10.83	10.90	12.36	26.78	0.62	0.63	Cas
30	3	0	46	60	7	18	10.76	11.27	24.79	94.94	1.75	0.57	Cas
31	11	39	29	-57	22	4	10.34	11.09	11.73	332.29	-0.07	0.30	Cen
32	11	44	51	-62	19	5	10.16	10.70	10.19	339.12	1.24	0.68	Cen
33	11	46	2	-57	6	35	10.31	11.25	22.56	88.30	1.41	0.18	Cen
34	13	45	57	-60	9	1	10.76	11.04	15.91	63.42	0.34	0.35	Cen
35	20	5	40	79	51	26	10.01	10.23	28.84	2.84	0.73	0.73	Cep
36	22	34	53	62	8	49	10.05	10.58	20.09	67.78	1.55	0.45	Cep

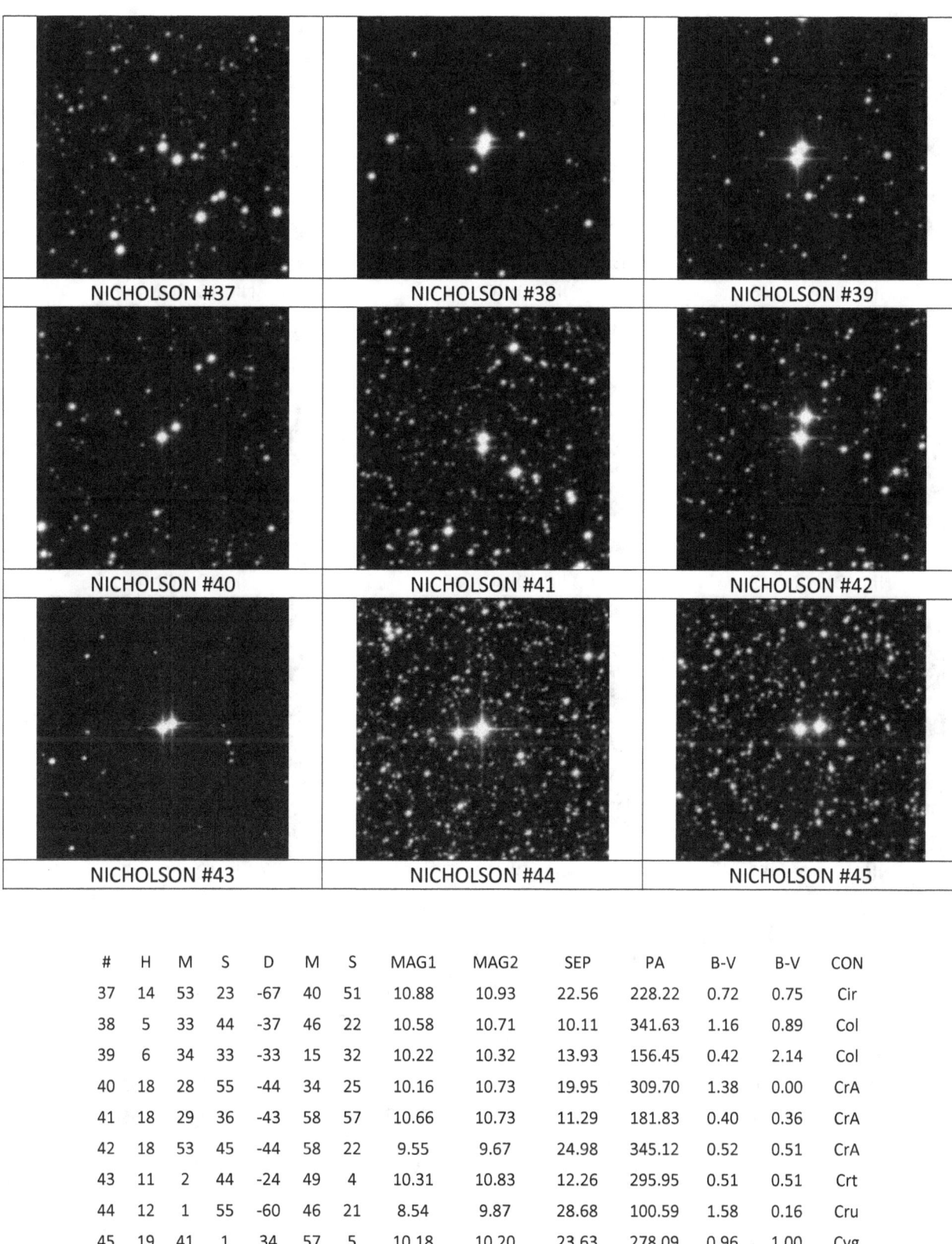

#	H	M	S	D	M	S	MAG1	MAG2	SEP	PA	B-V	B-V	CON
37	14	53	23	-67	40	51	10.88	10.93	22.56	228.22	0.72	0.75	Cir
38	5	33	44	-37	46	22	10.58	10.71	10.11	341.63	1.16	0.89	Col
39	6	34	33	-33	15	32	10.22	10.32	13.93	156.45	0.42	2.14	Col
40	18	28	55	-44	34	25	10.16	10.73	19.95	309.70	1.38	0.00	CrA
41	18	29	36	-43	58	57	10.66	10.73	11.29	181.83	0.40	0.36	CrA
42	18	53	45	-44	58	22	9.55	9.67	24.98	345.12	0.52	0.51	CrA
43	11	2	44	-24	49	4	10.31	10.83	12.26	295.95	0.51	0.51	Crt
44	12	1	55	-60	46	21	8.54	9.87	28.68	100.59	1.58	0.16	Cru
45	19	41	1	34	57	5	10.18	10.20	23.63	278.09	0.96	1.00	Cyg

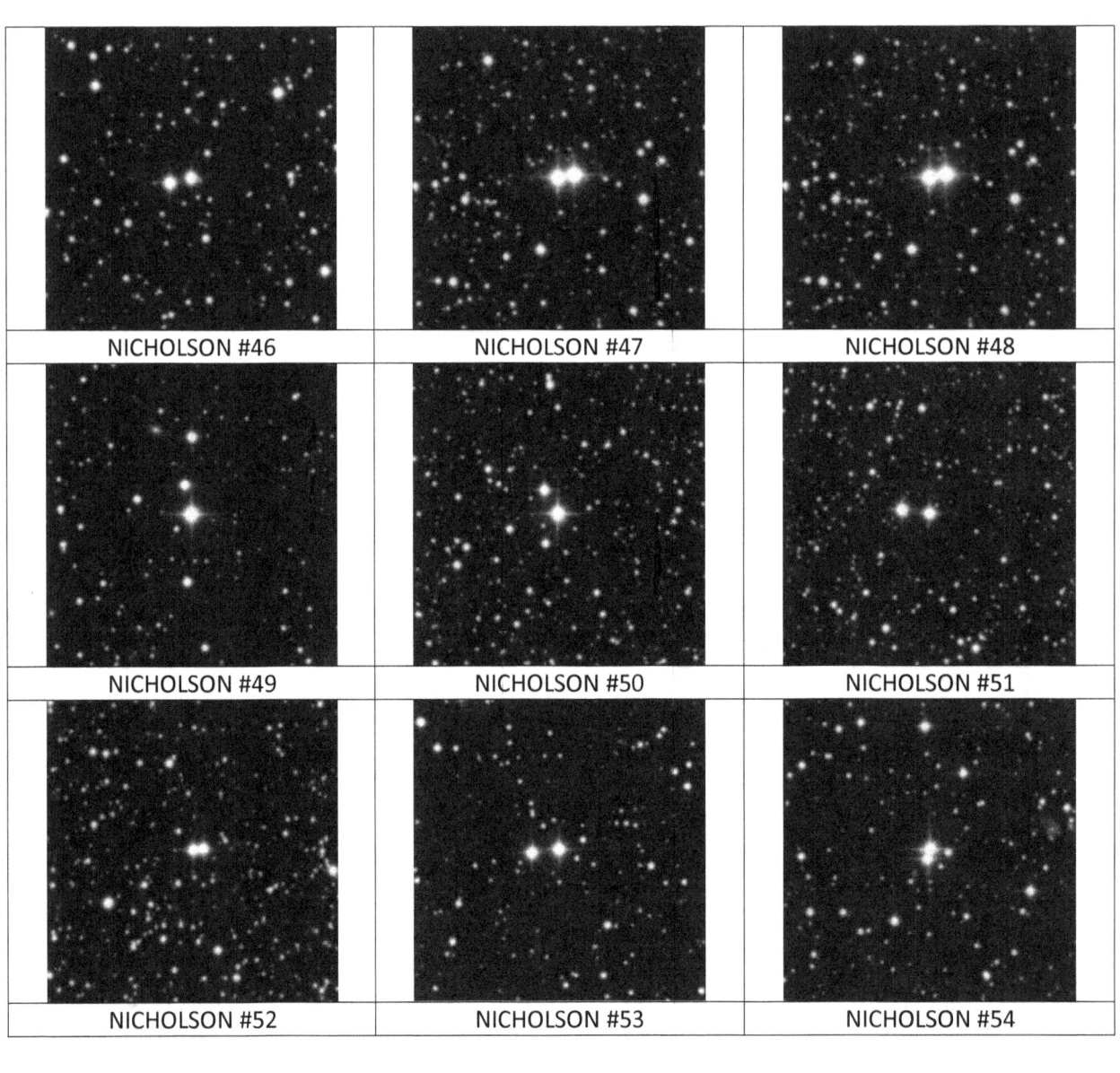

#	H	M	S	D	M	S	MAG1	MAG2	SEP	PA	B-V	B-V	CON
46	19	50	27	51	44	52	11.15	11.35	22.65	100.25	0.53	0.52	Cyg
47	20	3	39	43	3	17	11.14	11.22	20.32	130.58	0.81	0.81	Cyg
48	20	5	12	48	11	14	9.46	9.50	16.70	281.64	0.56	0.55	Cyg
49	20	17	53	59	47	4	9.50	9.93	29.83	13.09	0.24	1.27	Cyg
50	20	18	25	51	38	48	10.36	11.26	27.10	28.28	1.17	0.14	Cyg
51	20	20	11	48	15	29	10.49	10.50	28.34	80.64	0.47	0.61	Cyg
52	21	0	17	33	21	16	11.00	11.08	10.61	278.80	0.19	0.49	Cyg
53	21	40	49	31	17	33	10.41	10.84	28.17	98.48	0.36	0.36	Cyg
54	21	43	41	34	43	41	10.19	11.01	10.08	167.47	1.31	0.85	Cyg

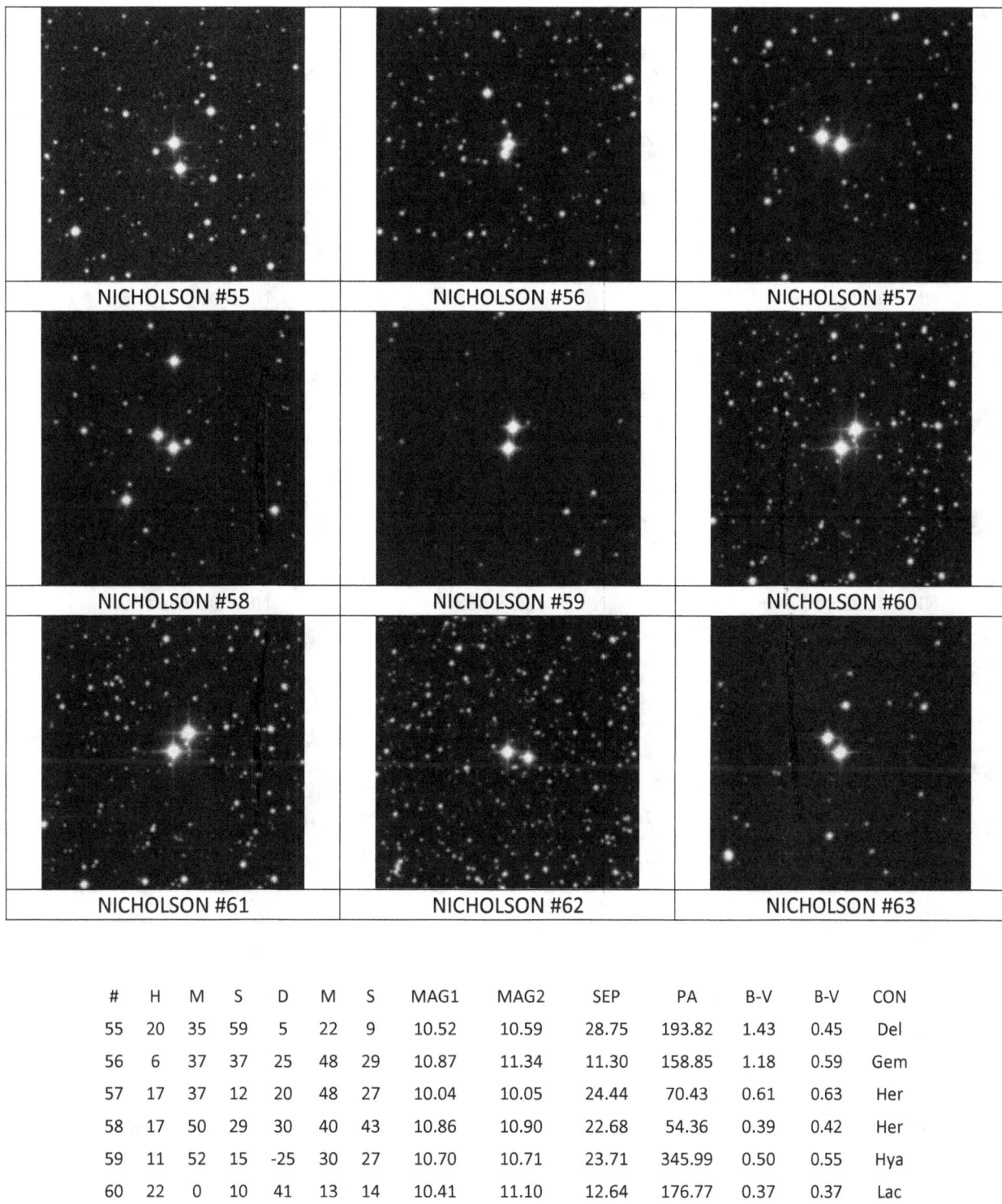

NICHOLSON #55	NICHOLSON #56	NICHOLSON #57
NICHOLSON #58	NICHOLSON #59	NICHOLSON #60
NICHOLSON #61	NICHOLSON #62	NICHOLSON #63

#	H	M	S	D	M	S	MAG1	MAG2	SEP	PA	B-V	B-V	CON
55	20	35	59	5	22	9	10.52	10.59	28.75	193.82	1.43	0.45	Del
56	6	37	37	25	48	29	10.87	11.34	11.30	158.85	1.18	0.59	Gem
57	17	37	12	20	48	27	10.04	10.05	24.44	70.43	0.61	0.63	Her
58	17	50	29	30	40	43	10.86	10.90	22.68	54.36	0.39	0.42	Her
59	11	52	15	-25	30	27	10.70	10.71	23.71	345.99	0.50	0.55	Hya
60	22	0	10	41	13	14	10.41	11.10	12.64	176.77	0.37	0.37	Lac
61	22	21	13	46	8	10	9.16	9.38	26.62	319.78	0.38	0.38	Lac
62	22	48	2	54	12	53	10.50	11.09	24.35	257.19	1.32	-0.15	Lac
63	5	29	32	-23	25	45	10.93	10.95	21.91	42.33	0.85	0.91	Lep

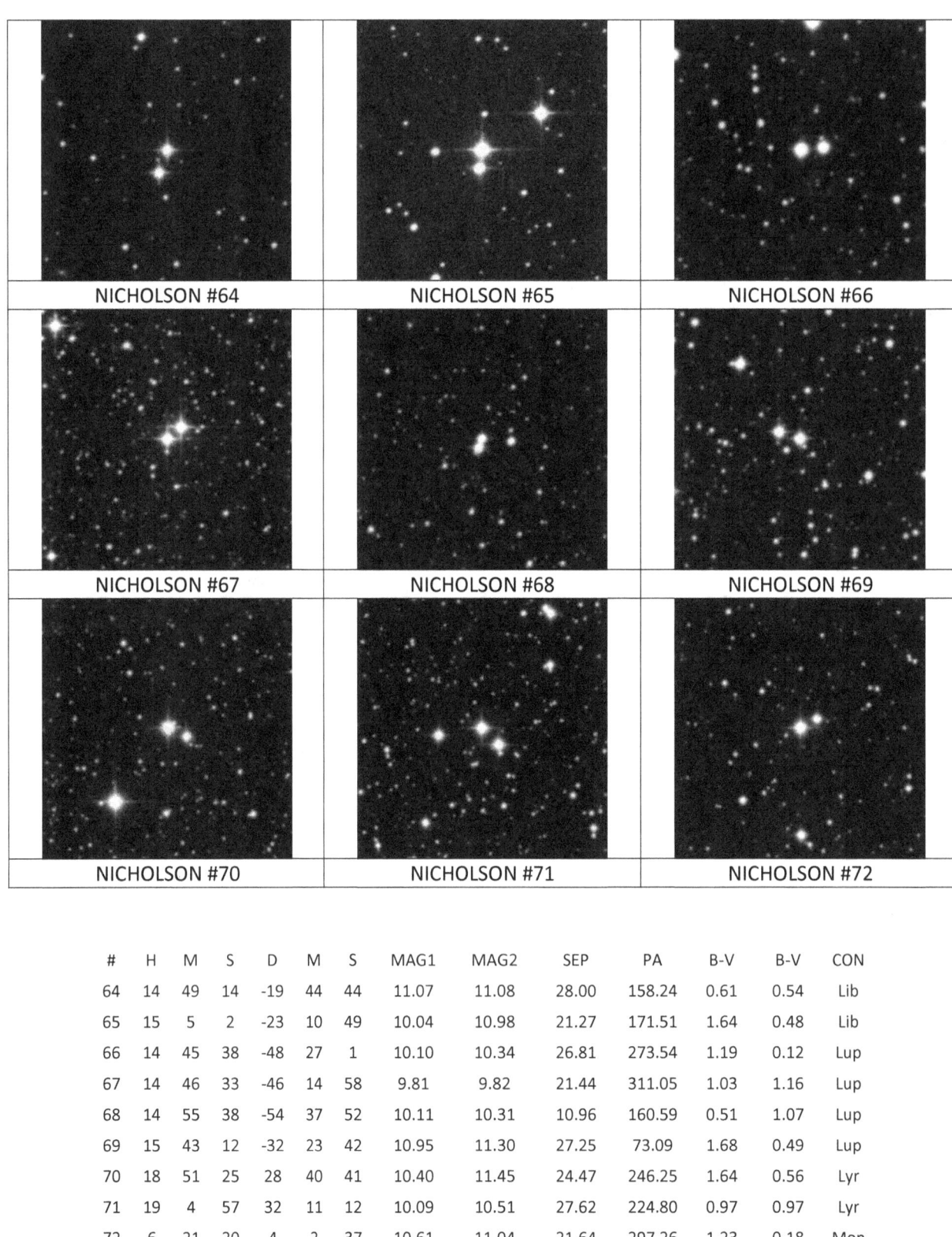

#	H	M	S	D	M	S	MAG1	MAG2	SEP	PA	B-V	B-V	CON
64	14	49	14	-19	44	44	11.07	11.08	28.00	158.24	0.61	0.54	Lib
65	15	5	2	-23	10	49	10.04	10.98	21.27	171.51	1.64	0.48	Lib
66	14	45	38	-48	27	1	10.10	10.34	26.81	273.54	1.19	0.12	Lup
67	14	46	33	-46	14	58	9.81	9.82	21.44	311.05	1.03	1.16	Lup
68	14	55	38	-54	37	52	10.11	10.31	10.96	160.59	0.51	1.07	Lup
69	15	43	12	-32	23	42	10.95	11.30	27.25	73.09	1.68	0.49	Lup
70	18	51	25	28	40	41	10.40	11.45	24.47	246.25	1.64	0.56	Lyr
71	19	4	57	32	11	12	10.09	10.51	27.62	224.80	0.97	0.97	Lyr
72	6	21	20	4	2	37	10.61	11.04	21.64	297.26	1.23	0.18	Mon

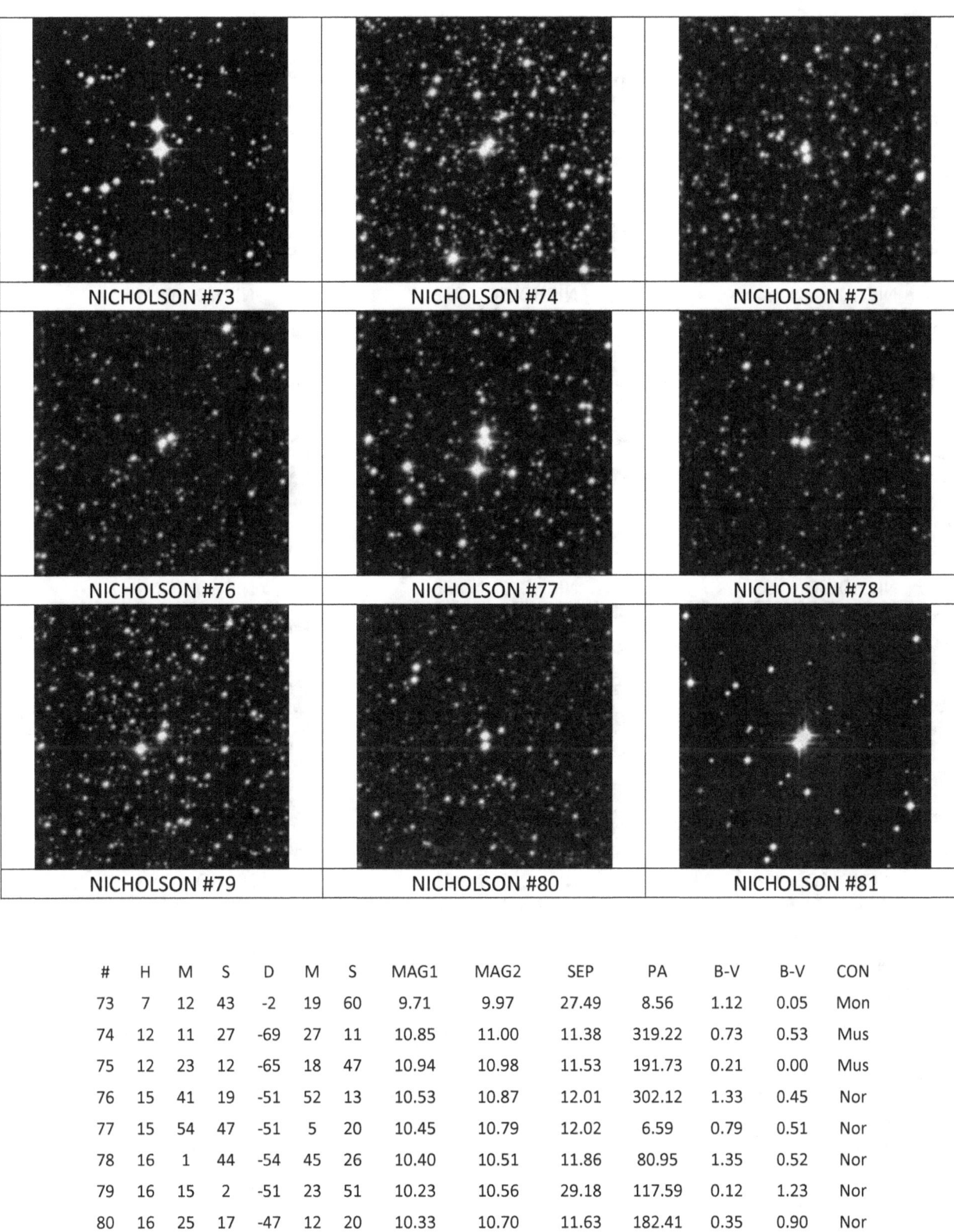

NICHOLSON #73	NICHOLSON #74	NICHOLSON #75
NICHOLSON #76	NICHOLSON #77	NICHOLSON #78
NICHOLSON #79	NICHOLSON #80	NICHOLSON #81

#	H	M	S	D	M	S	MAG1	MAG2	SEP	PA	B-V	B-V	CON
73	7	12	43	-2	19	60	9.71	9.97	27.49	8.56	1.12	0.05	Mon
74	12	11	27	-69	27	11	10.85	11.00	11.38	319.22	0.73	0.53	Mus
75	12	23	12	-65	18	47	10.94	10.98	11.53	191.73	0.21	0.00	Mus
76	15	41	19	-51	52	13	10.53	10.87	12.01	302.12	1.33	0.45	Nor
77	15	54	47	-51	5	20	10.45	10.79	12.02	6.59	0.79	0.51	Nor
78	16	1	44	-54	45	26	10.40	10.51	11.86	80.95	1.35	0.52	Nor
79	16	15	2	-51	23	51	10.23	10.56	29.18	117.59	0.12	1.23	Nor
80	16	25	17	-47	12	20	10.33	10.70	11.63	182.41	0.35	0.90	Nor
81	5	25	47	-9	16	43	10.70	10.92	11.67	139.88	0.70	0.58	Ori

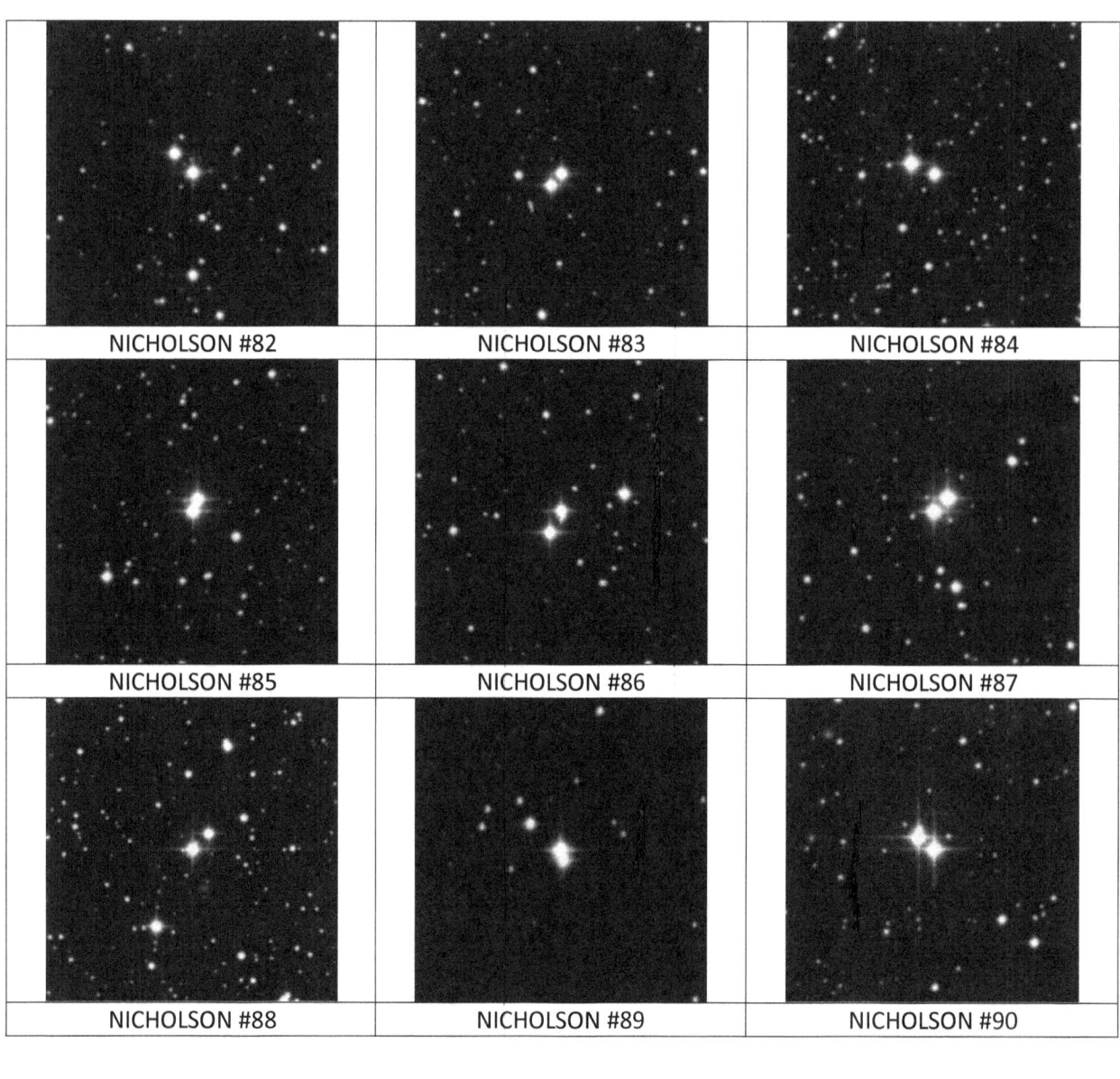

#	H	M	S	D	M	S	MAG1	MAG2	SEP	PA	B-V	B-V	CON
82	5	33	49	9	57	60	10.53	10.61	26.77	44.76	0.56	0.56	Ori
83	5	34	50	10	34	50	10.57	10.67	15.48	138.22	0.48	0.50	Ori
84	5	59	27	22	34	40	10.38	10.51	25.56	66.71	0.29	1.65	Ori
85	6	7	6	2	52	30	9.68	9.88	12.24	338.33	0.36	0.52	Ori
86	21	35	24	9	29	18	10.06	10.33	24.29	150.89	-0.05	0.95	Peg
87	22	22	30	28	14	25	9.87	9.95	19.43	314.80	0.64	0.65	Peg
88	3	9	18	48	17	39	10.13	11.28	22.75	314.98	0.58	0.58	Per
89	4	59	35	-48	26	42	10.39	11.11	10.77	197.86	0.39	0.60	Pic
90	6	48	28	-62	12	15	9.49	9.51	20.07	55.54	0.33	0.39	Pic

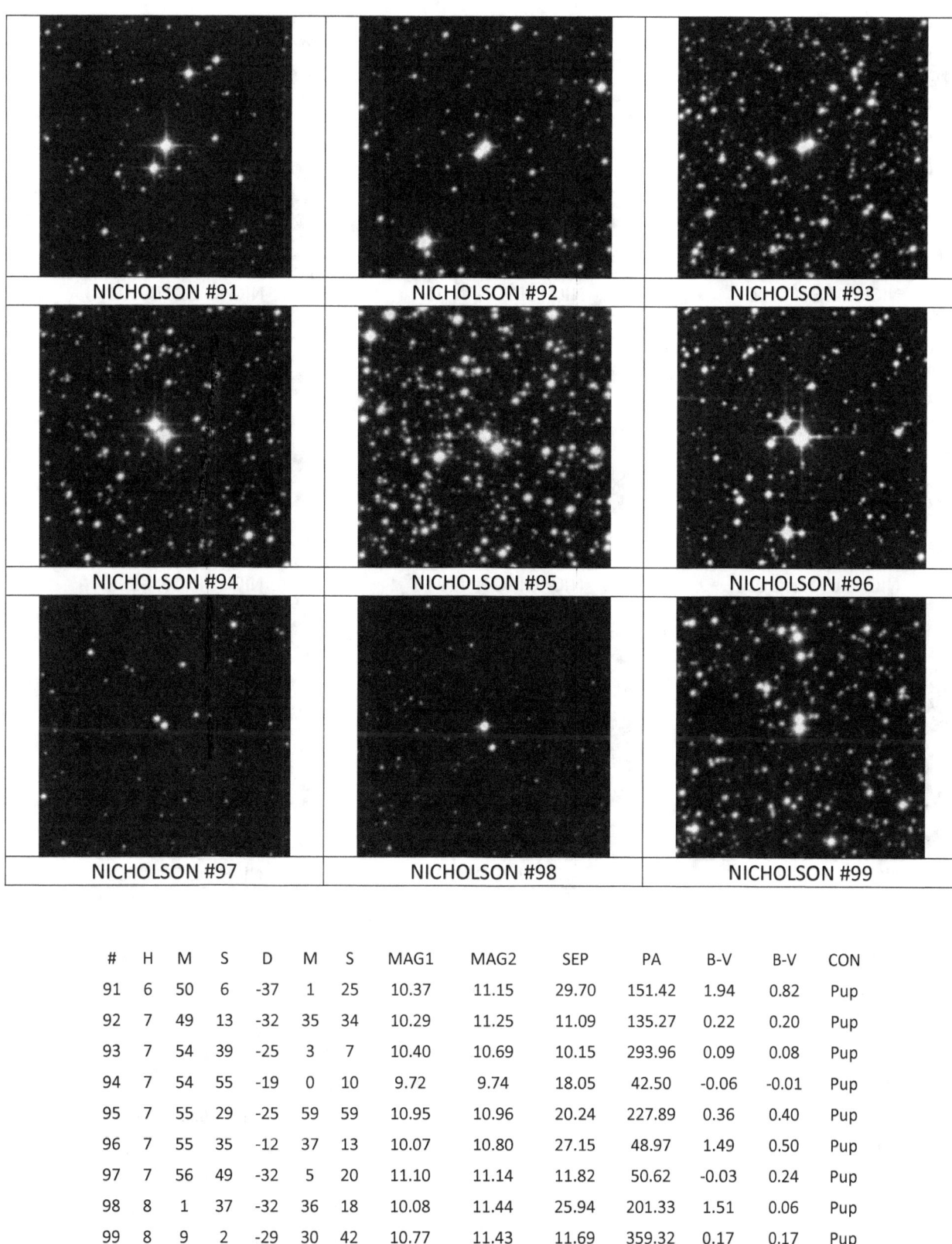

#	H	M	S	D	M	S	MAG1	MAG2	SEP	PA	B-V	B-V	CON
91	6	50	6	-37	1	25	10.37	11.15	29.70	151.42	1.94	0.82	Pup
92	7	49	13	-32	35	34	10.29	11.25	11.09	135.27	0.22	0.20	Pup
93	7	54	39	-25	3	7	10.40	10.69	10.15	293.96	0.09	0.08	Pup
94	7	54	55	-19	0	10	9.72	9.74	18.05	42.50	-0.06	-0.01	Pup
95	7	55	29	-25	59	59	10.95	10.96	20.24	227.89	0.36	0.40	Pup
96	7	55	35	-12	37	13	10.07	10.80	27.15	48.97	1.49	0.50	Pup
97	7	56	49	-32	5	20	11.10	11.14	11.82	50.62	-0.03	0.24	Pup
98	8	1	37	-32	36	18	10.08	11.44	25.94	201.33	1.51	0.06	Pup
99	8	9	2	-29	30	42	10.77	11.43	11.69	359.32	0.17	0.17	Pup

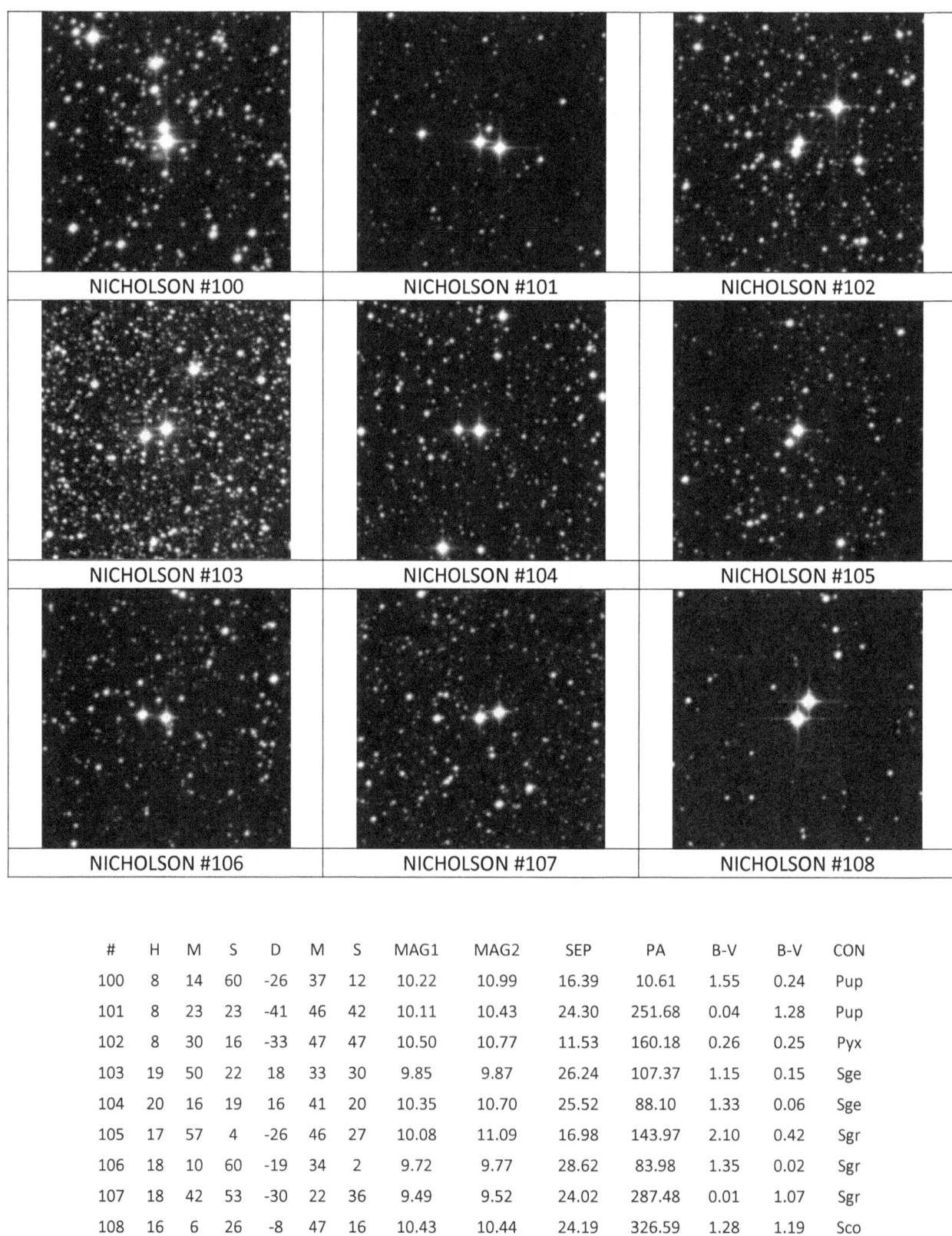

#	H	M	S	D	M	S	MAG1	MAG2	SEP	PA	B-V	B-V	CON
100	8	14	60	-26	37	12	10.22	10.99	16.39	10.61	1.55	0.24	Pup
101	8	23	23	-41	46	42	10.11	10.43	24.30	251.68	0.04	1.28	Pup
102	8	30	16	-33	47	47	10.50	10.77	11.53	160.18	0.26	0.25	Pyx
103	19	50	22	18	33	30	9.85	9.87	26.24	107.37	1.15	0.15	Sge
104	20	16	19	16	41	20	10.35	10.70	25.52	88.10	1.33	0.06	Sge
105	17	57	4	-26	46	27	10.08	11.09	16.98	143.97	2.10	0.42	Sgr
106	18	10	60	-19	34	2	9.72	9.77	28.62	83.98	1.35	0.02	Sgr
107	18	42	53	-30	22	36	9.49	9.52	24.02	287.48	0.01	1.07	Sgr
108	16	6	26	-8	47	16	10.43	10.44	24.19	326.59	1.28	1.19	Sco

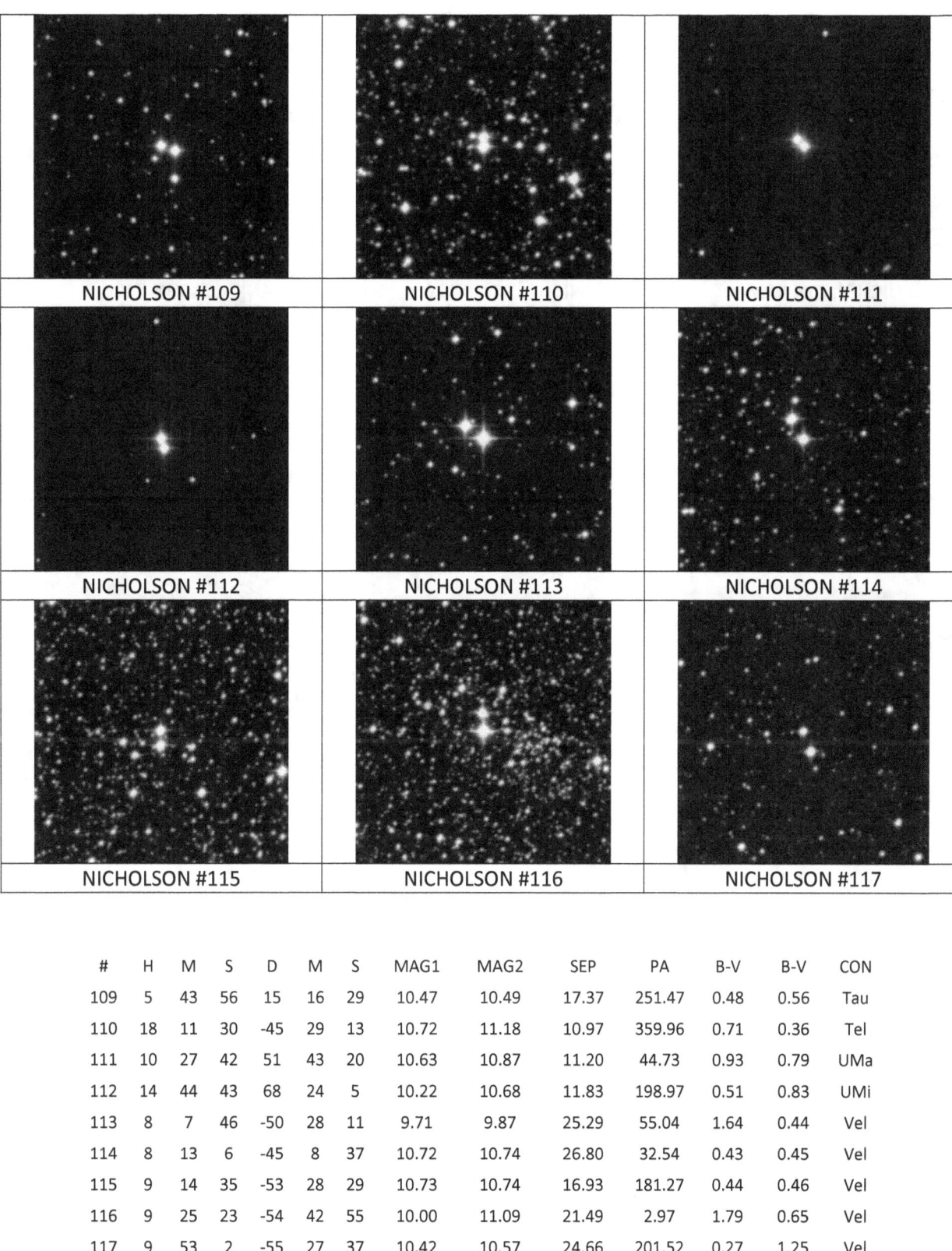

#	H	M	S	D	M	S	MAG1	MAG2	SEP	PA	B-V	B-V	CON
109	5	43	56	15	16	29	10.47	10.49	17.37	251.47	0.48	0.56	Tau
110	18	11	30	-45	29	13	10.72	11.18	10.97	359.96	0.71	0.36	Tel
111	10	27	42	51	43	20	10.63	10.87	11.20	44.73	0.93	0.79	UMa
112	14	44	43	68	24	5	10.22	10.68	11.83	198.97	0.51	0.83	UMi
113	8	7	46	-50	28	11	9.71	9.87	25.29	55.04	1.64	0.44	Vel
114	8	13	6	-45	8	37	10.72	10.74	26.80	32.54	0.43	0.45	Vel
115	9	14	35	-53	28	29	10.73	10.74	16.93	181.27	0.44	0.46	Vel
116	9	25	23	-54	42	55	10.00	11.09	21.49	2.97	1.79	0.65	Vel
117	9	53	2	-55	27	37	10.42	10.57	24.66	201.52	0.27	1.25	Vel

| NICHOLSON #118 | NICHOLSON #119 | NICHOLSON #120 |

#	H	M	S	D	M	S	MAG1	MAG2	SEP	PA	B-V	B-V	CON
118	9	56	48	-53	33	45	10.16	10.20	11.72	302.12	0.96	0.42	Vel
119	10	44	13	-50	37	47	10.99	11.00	11.59	174.88	0.39	0.56	Vel
120	20	52	34	28	30	10	10.21	10.31	20.44	172.59	1.29	1.27	Vul

Nicholson #120 to #240

Careful measurement over many years reveals that all stars are moving independently through space and this results in a slow change in their position relative to the earth. This is called "proper motion" and an animation of this phenomenon is available at:-

http://www.martin-nicholson.info/cpm/cpmmaster.htm

Proper motion is a vector, that is it has both a magnitude and a direction. The magnitude has units of arc seconds per year and the direction is expressed in degrees with 0 degrees being north, 90 degrees being east and so on. Most catalogues present proper motion information in the form of the magnitude of the motion in both right ascension and in declination since these are at right angles to each other.

For a pair to be considered a common proper motion binary star the two components would be expected to show very similar proper motion. Reliable and up-to-date results are not available for most stars but where such information exists it is clear that there are many previously unlisted binary stars just waiting to be observed.

These 120 objects were selected from the large number of common proper motion binary stars that were first described in my book "Binary star discoveries in the URAT1 catalog". This book is available from Amazon.com and Amazon.co.uk.

Explanation of the data tables:

1. # – The Nicholson number of the confirmed common proper motion star.
2. RA – The right ascension of the primary star in hours, minutes and seconds.
3. DEC – The declination of the primary star in degrees, minutes and seconds.
4. MAG1 and MAG2 – The magnitudes of the primary and secondary stars.
5. SEP – The distance between the stars in arc seconds.
6. CON – The abbreviated name of the constellation in which the double star can be found.

The finder charts have dimensions of 48 by 48 arc seconds.

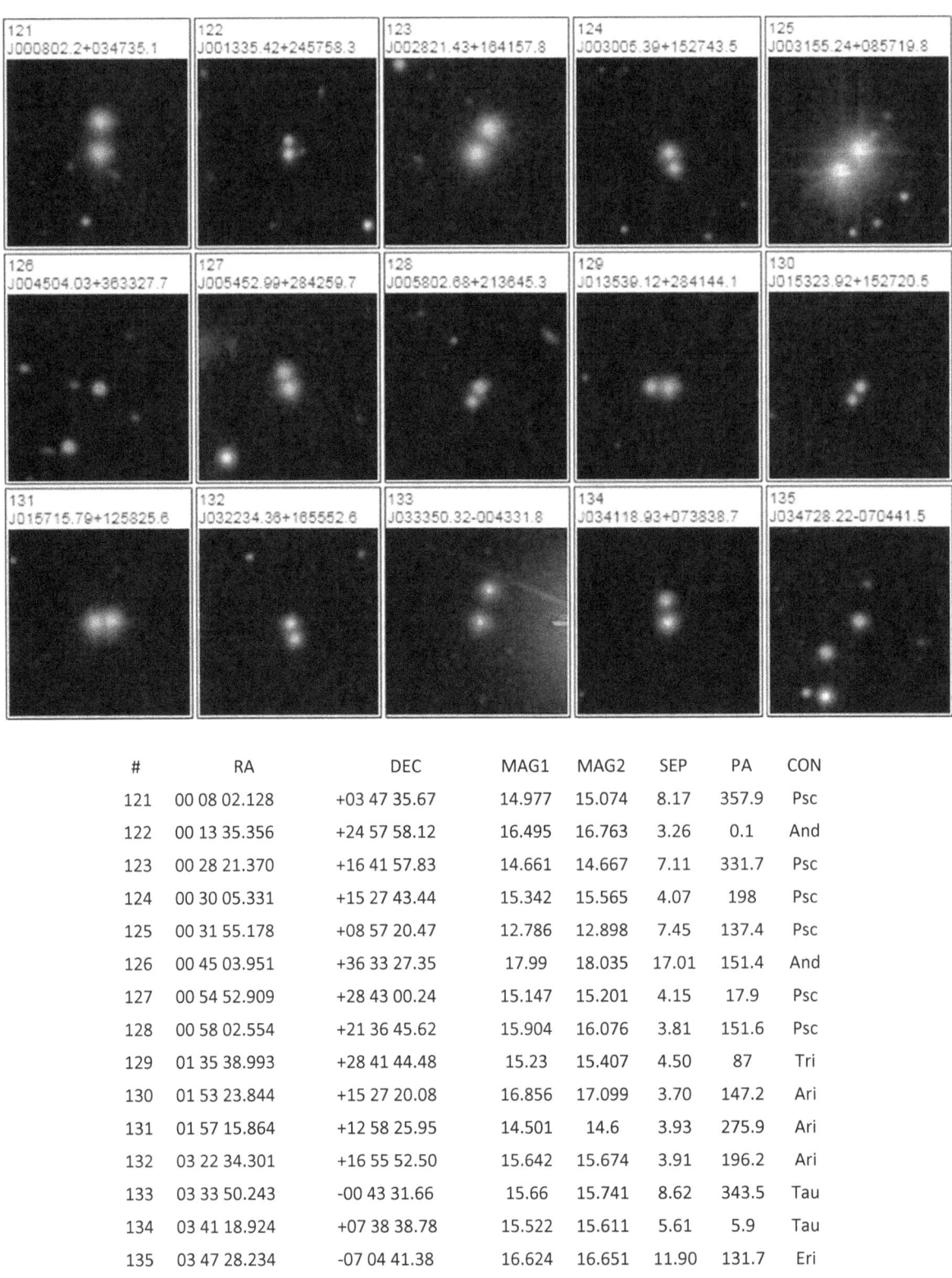

#	RA	DEC	MAG1	MAG2	SEP	PA	CON
121	00 08 02.128	+03 47 35.67	14.977	15.074	8.17	357.9	Psc
122	00 13 35.356	+24 57 58.12	16.495	16.763	3.26	0.1	And
123	00 28 21.370	+16 41 57.83	14.661	14.667	7.11	331.7	Psc
124	00 30 05.331	+15 27 43.44	15.342	15.565	4.07	198	Psc
125	00 31 55.178	+08 57 20.47	12.786	12.898	7.45	137.4	Psc
126	00 45 03.951	+36 33 27.35	17.99	18.035	17.01	151.4	And
127	00 54 52.909	+28 43 00.24	15.147	15.201	4.15	17.9	Psc
128	00 58 02.554	+21 36 45.62	15.904	16.076	3.81	151.6	Psc
129	01 35 38.993	+28 41 44.48	15.23	15.407	4.50	87	Tri
130	01 53 23.844	+15 27 20.08	16.856	17.099	3.70	147.2	Ari
131	01 57 15.864	+12 58 25.95	14.501	14.6	3.93	275.9	Ari
132	03 22 34.301	+16 55 52.50	15.642	15.674	3.91	196.2	Ari
133	03 33 50.243	-00 43 31.66	15.66	15.741	8.62	343.5	Tau
134	03 41 18.924	+07 38 38.78	15.522	15.611	5.61	5.9	Tau
135	03 47 28.234	-07 04 41.38	16.624	16.651	11.90	131.7	Eri

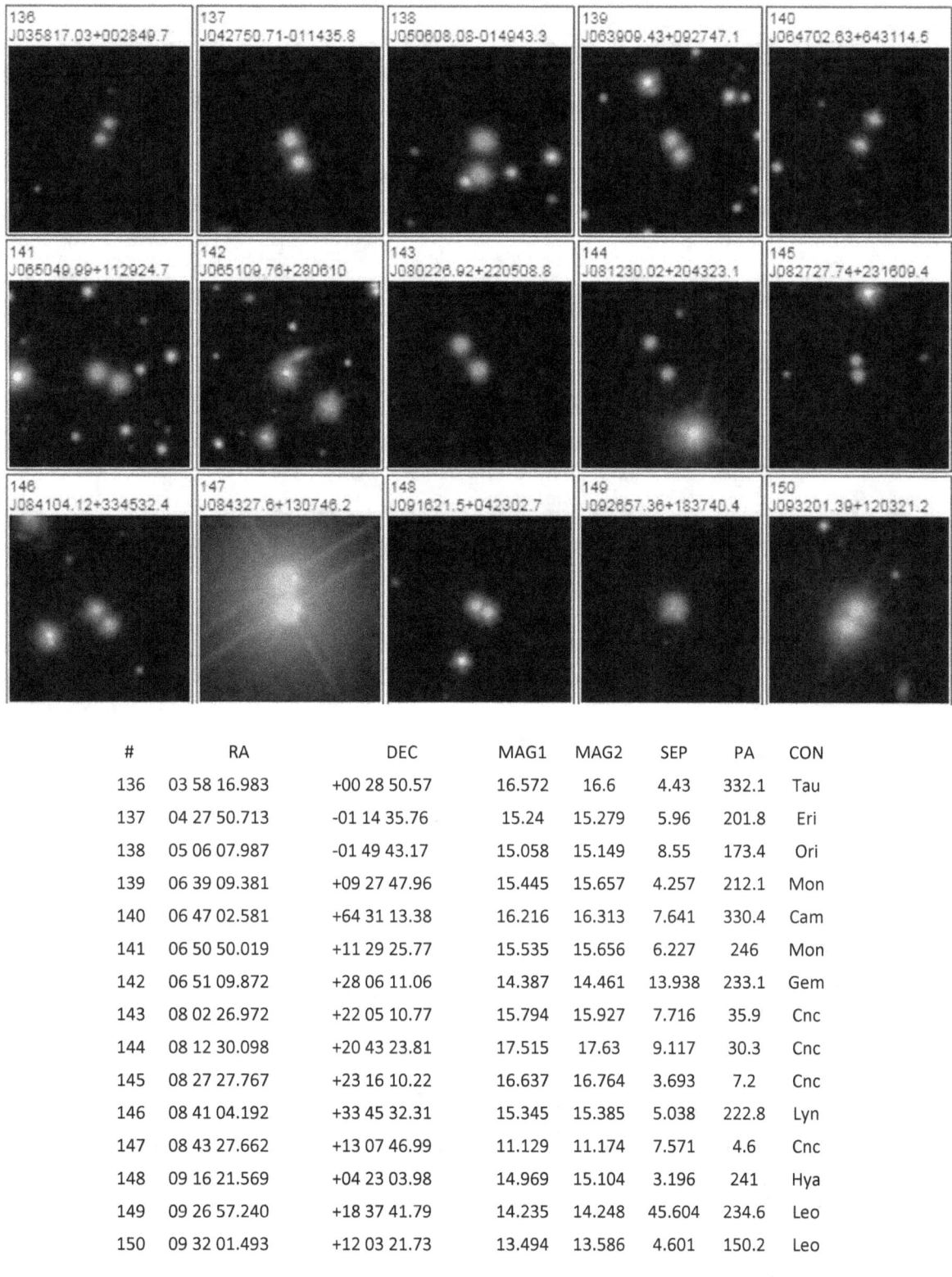

#	RA	DEC	MAG1	MAG2	SEP	PA	CON
136	03 58 16.983	+00 28 50.57	16.572	16.6	4.43	332.1	Tau
137	04 27 50.713	-01 14 35.76	15.24	15.279	5.96	201.8	Eri
138	05 06 07.987	-01 49 43.17	15.058	15.149	8.55	173.4	Ori
139	06 39 09.381	+09 27 47.96	15.445	15.657	4.257	212.1	Mon
140	06 47 02.581	+64 31 13.38	16.216	16.313	7.641	330.4	Cam
141	06 50 50.019	+11 29 25.77	15.535	15.656	6.227	246	Mon
142	06 51 09.872	+28 06 11.06	14.387	14.461	13.938	233.1	Gem
143	08 02 26.972	+22 05 10.77	15.794	15.927	7.716	35.9	Cnc
144	08 12 30.098	+20 43 23.81	17.515	17.63	9.117	30.3	Cnc
145	08 27 27.767	+23 16 10.22	16.637	16.764	3.693	7.2	Cnc
146	08 41 04.192	+33 45 32.31	15.345	15.385	5.038	222.8	Lyn
147	08 43 27.662	+13 07 46.99	11.129	11.174	7.571	4.6	Cnc
148	09 16 21.569	+04 23 03.98	14.969	15.104	3.196	241	Hya
149	09 26 57.240	+18 37 41.79	14.235	14.248	45.604	234.6	Leo
150	09 32 01.493	+12 03 21.73	13.494	13.586	4.601	150.2	Leo

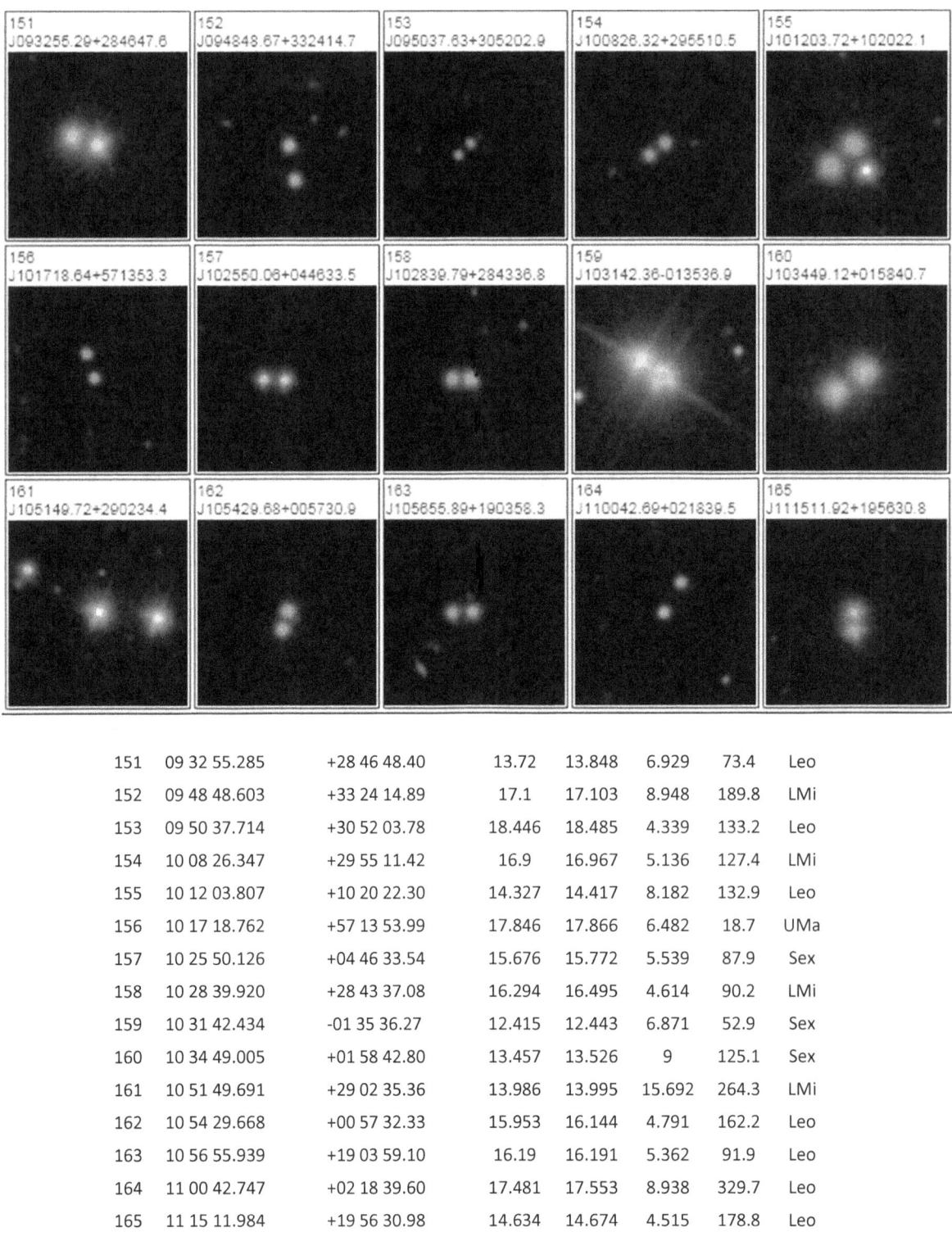

151	09 32 55.285	+28 46 48.40	13.72	13.848	6.929	73.4	Leo
152	09 48 48.603	+33 24 14.89	17.1	17.103	8.948	189.8	LMi
153	09 50 37.714	+30 52 03.78	18.446	18.485	4.339	133.2	Leo
154	10 08 26.347	+29 55 11.42	16.9	16.967	5.136	127.4	LMi
155	10 12 03.807	+10 20 22.30	14.327	14.417	8.182	132.9	Leo
156	10 17 18.762	+57 13 53.99	17.846	17.866	6.482	18.7	UMa
157	10 25 50.126	+04 46 33.54	15.676	15.772	5.539	87.9	Sex
158	10 28 39.920	+28 43 37.08	16.294	16.495	4.614	90.2	LMi
159	10 31 42.434	-01 35 36.27	12.415	12.443	6.871	52.9	Sex
160	10 34 49.005	+01 58 42.80	13.457	13.526	9	125.1	Sex
161	10 51 49.691	+29 02 35.36	13.986	13.995	15.692	264.3	LMi
162	10 54 29.668	+00 57 32.33	15.953	16.144	4.791	162.2	Leo
163	10 56 55.939	+19 03 59.10	16.19	16.191	5.362	91.9	Leo
164	11 00 42.747	+02 18 39.60	17.481	17.553	8.938	329.7	Leo
165	11 15 11.984	+19 56 30.98	14.634	14.674	4.515	178.8	Leo

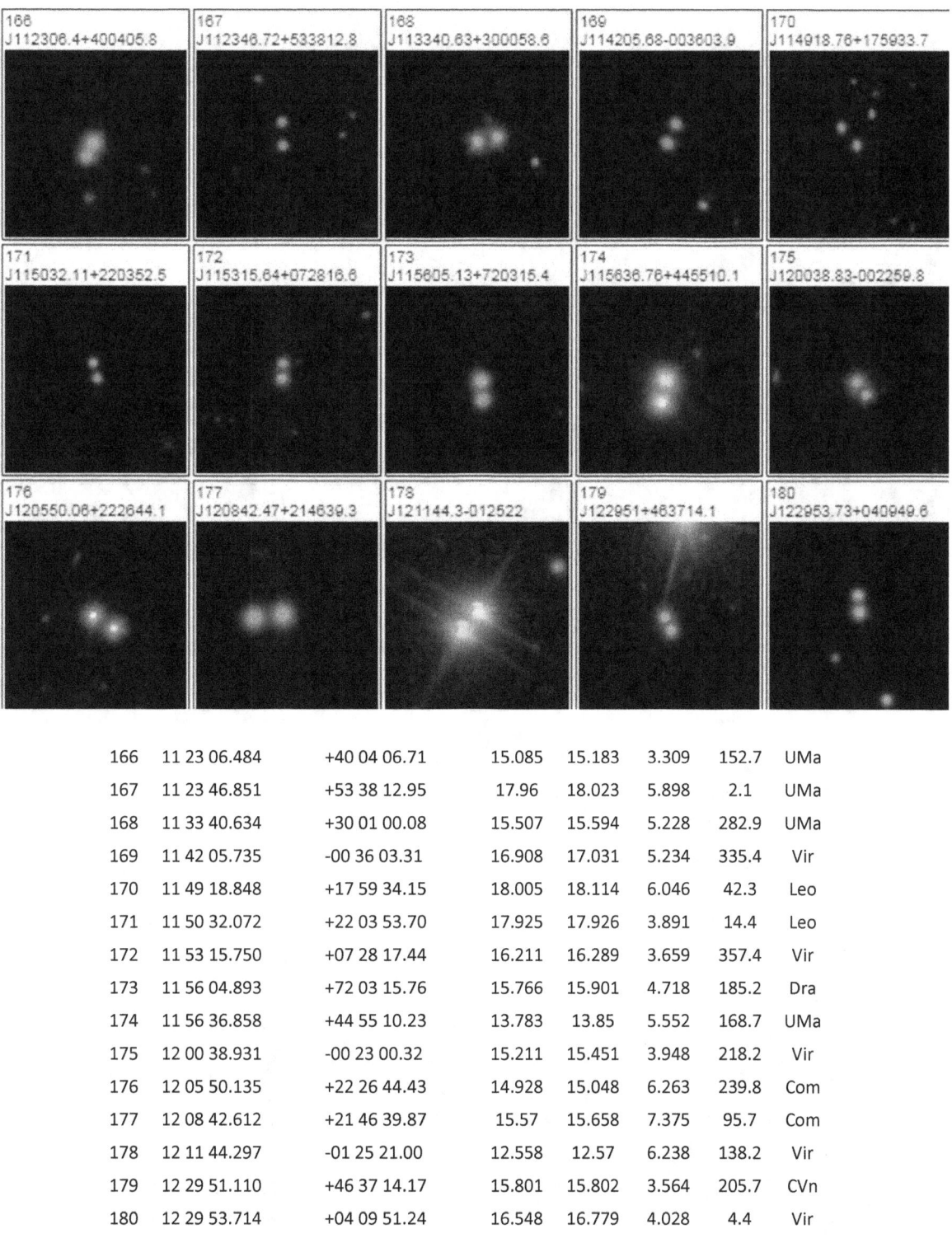

166	11 23 06.484	+40 04 06.71	15.085	15.183	3.309	152.7	UMa
167	11 23 46.851	+53 38 12.95	17.96	18.023	5.898	2.1	UMa
168	11 33 40.634	+30 01 00.08	15.507	15.594	5.228	282.9	UMa
169	11 42 05.735	-00 36 03.31	16.908	17.031	5.234	335.4	Vir
170	11 49 18.848	+17 59 34.15	18.005	18.114	6.046	42.3	Leo
171	11 50 32.072	+22 03 53.70	17.925	17.926	3.891	14.4	Leo
172	11 53 15.750	+07 28 17.44	16.211	16.289	3.659	357.4	Vir
173	11 56 04.893	+72 03 15.76	15.766	15.901	4.718	185.2	Dra
174	11 56 36.858	+44 55 10.23	13.783	13.85	5.552	168.7	UMa
175	12 00 38.931	-00 23 00.32	15.211	15.451	3.948	218.2	Vir
176	12 05 50.135	+22 26 44.43	14.928	15.048	6.263	239.8	Com
177	12 08 42.612	+21 46 39.87	15.57	15.658	7.375	95.7	Com
178	12 11 44.297	-01 25 21.00	12.558	12.57	6.238	138.2	Vir
179	12 29 51.110	+46 37 14.17	15.801	15.802	3.564	205.7	CVn
180	12 29 53.714	+04 09 51.24	16.548	16.779	4.028	4.4	Vir

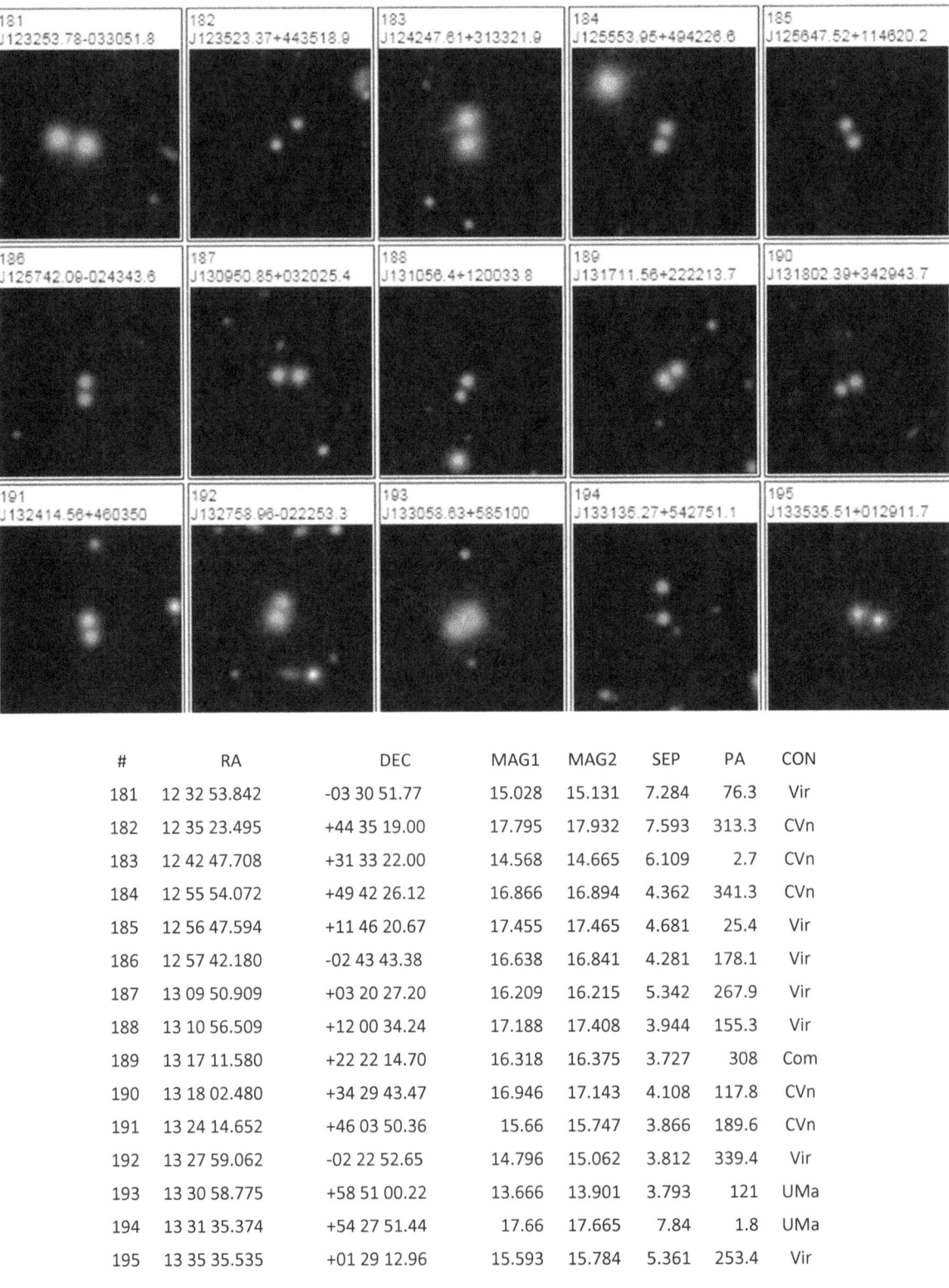

#	RA	DEC	MAG1	MAG2	SEP	PA	CON
181	12 32 53.842	-03 30 51.77	15.028	15.131	7.284	76.3	Vir
182	12 35 23.495	+44 35 19.00	17.795	17.932	7.593	313.3	CVn
183	12 42 47.708	+31 33 22.00	14.568	14.665	6.109	2.7	CVn
184	12 55 54.072	+49 42 26.12	16.866	16.894	4.362	341.3	CVn
185	12 56 47.594	+11 46 20.67	17.455	17.465	4.681	25.4	Vir
186	12 57 42.180	-02 43 43.38	16.638	16.841	4.281	178.1	Vir
187	13 09 50.909	+03 20 27.20	16.209	16.215	5.342	267.9	Vir
188	13 10 56.509	+12 00 34.24	17.188	17.408	3.944	155.3	Vir
189	13 17 11.580	+22 22 14.70	16.318	16.375	3.727	308	Com
190	13 18 02.480	+34 29 43.47	16.946	17.143	4.108	117.8	CVn
191	13 24 14.652	+46 03 50.36	15.66	15.747	3.866	189.6	CVn
192	13 27 59.062	-02 22 52.65	14.796	15.062	3.812	339.4	Vir
193	13 30 58.775	+58 51 00.22	13.666	13.901	3.793	121	UMa
194	13 31 35.374	+54 27 51.44	17.66	17.665	7.84	1.8	UMa
195	13 35 35.535	+01 29 12.96	15.593	15.784	5.361	253.4	Vir

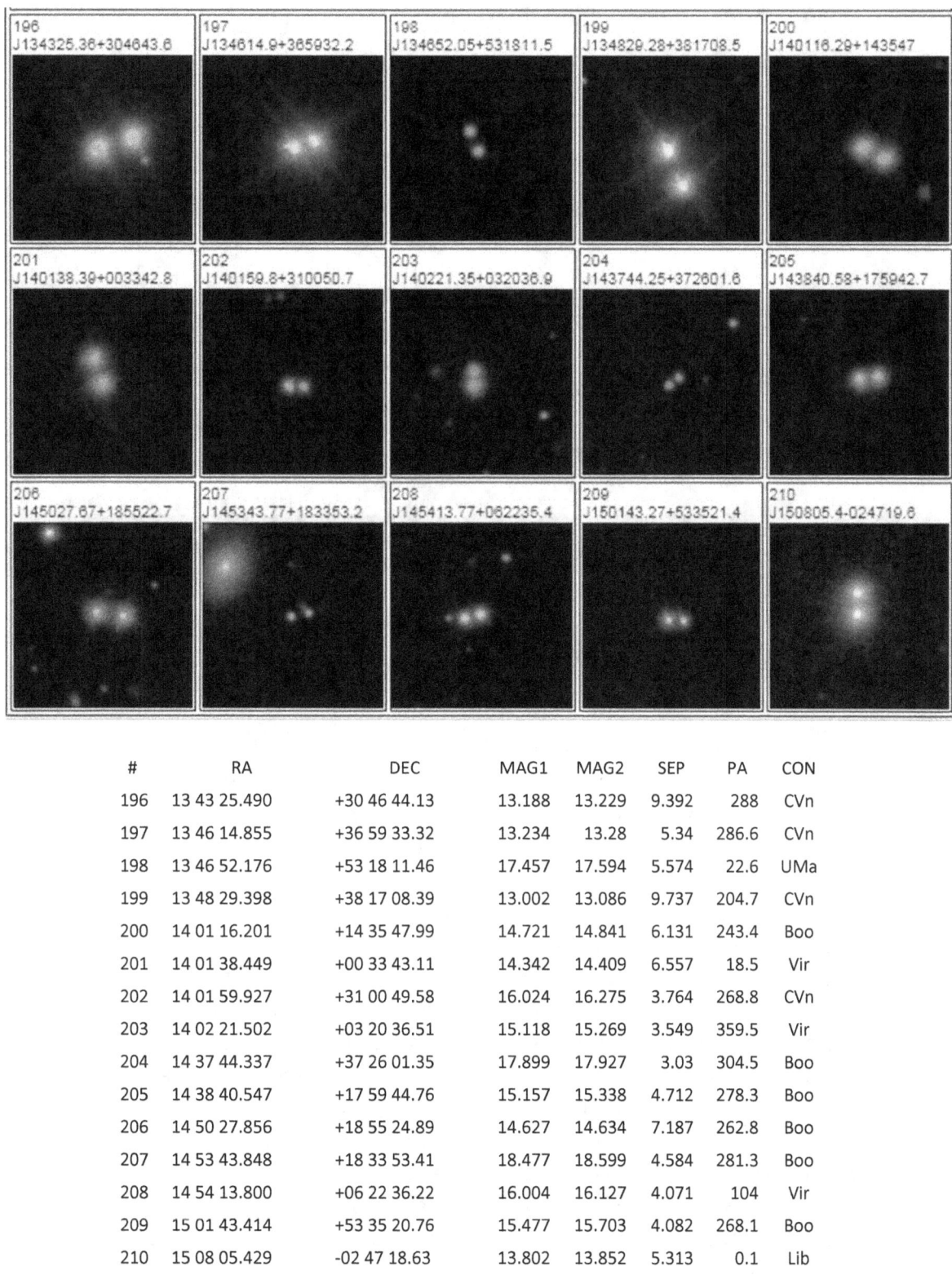

#	RA	DEC	MAG1	MAG2	SEP	PA	CON
196	13 43 25.490	+30 46 44.13	13.188	13.229	9.392	288	CVn
197	13 46 14.855	+36 59 33.32	13.234	13.28	5.34	286.6	CVn
198	13 46 52.176	+53 18 11.46	17.457	17.594	5.574	22.6	UMa
199	13 48 29.398	+38 17 08.39	13.002	13.086	9.737	204.7	CVn
200	14 01 16.201	+14 35 47.99	14.721	14.841	6.131	243.4	Boo
201	14 01 38.449	+00 33 43.11	14.342	14.409	6.557	18.5	Vir
202	14 01 59.927	+31 00 49.58	16.024	16.275	3.764	268.8	CVn
203	14 02 21.502	+03 20 36.51	15.118	15.269	3.549	359.5	Vir
204	14 37 44.337	+37 26 01.35	17.899	17.927	3.03	304.5	Boo
205	14 38 40.547	+17 59 44.76	15.157	15.338	4.712	278.3	Boo
206	14 50 27.856	+18 55 24.89	14.627	14.634	7.187	262.8	Boo
207	14 53 43.848	+18 33 53.41	18.477	18.599	4.584	281.3	Boo
208	14 54 13.800	+06 22 36.22	16.004	16.127	4.071	104	Vir
209	15 01 43.414	+53 35 20.76	15.477	15.703	4.082	268.1	Boo
210	15 08 05.429	-02 47 18.63	13.802	13.852	5.313	0.1	Lib

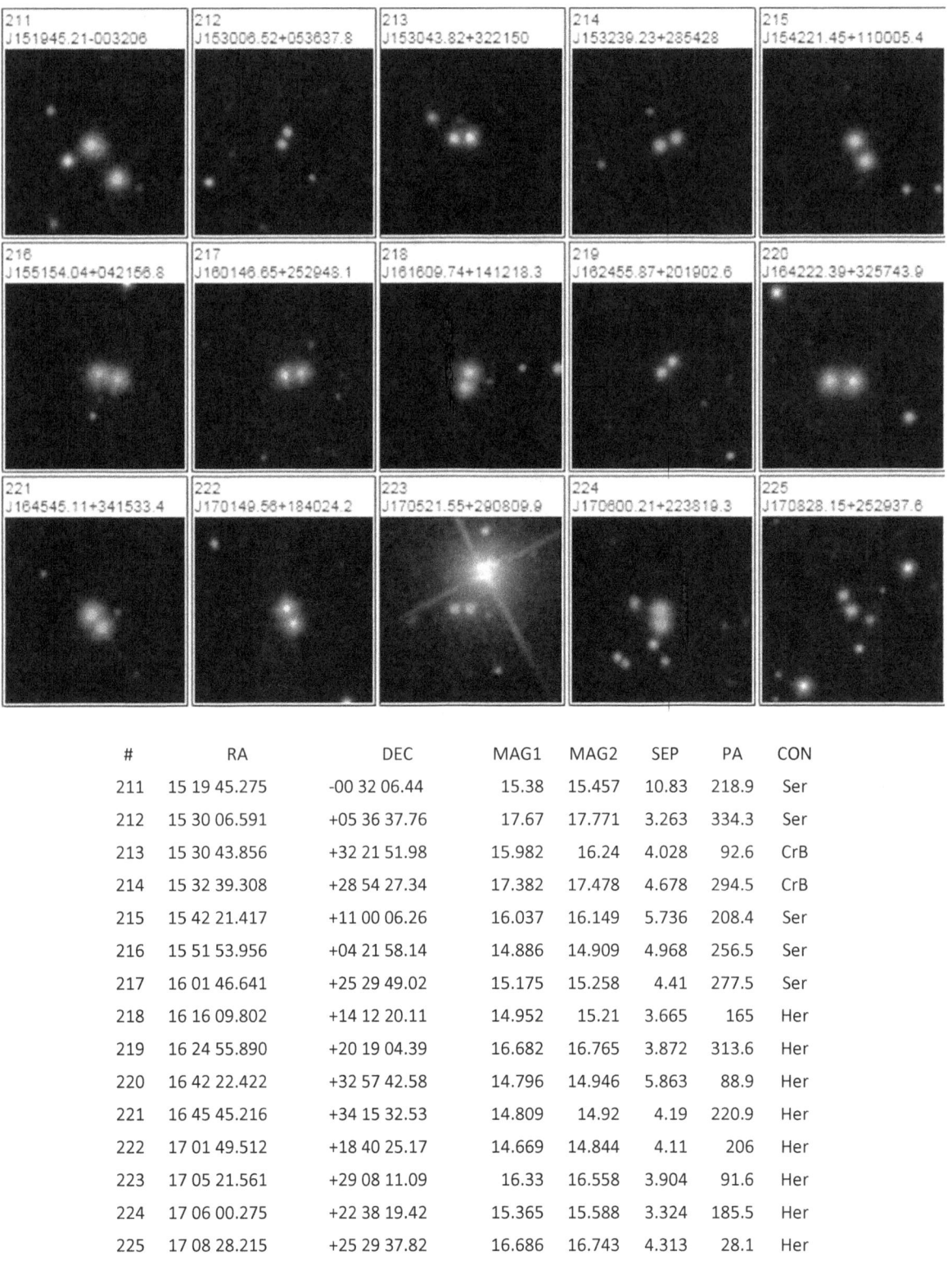

#	RA	DEC	MAG1	MAG2	SEP	PA	CON
211	15 19 45.275	-00 32 06.44	15.38	15.457	10.83	218.9	Ser
212	15 30 06.591	+05 36 37.76	17.67	17.771	3.263	334.3	Ser
213	15 30 43.856	+32 21 51.98	15.982	16.24	4.028	92.6	CrB
214	15 32 39.308	+28 54 27.34	17.382	17.478	4.678	294.5	CrB
215	15 42 21.417	+11 00 06.26	16.037	16.149	5.736	208.4	Ser
216	15 51 53.956	+04 21 58.14	14.886	14.909	4.968	256.5	Ser
217	16 01 46.641	+25 29 49.02	15.175	15.258	4.41	277.5	Ser
218	16 16 09.802	+14 12 20.11	14.952	15.21	3.665	165	Her
219	16 24 55.890	+20 19 04.39	16.682	16.765	3.872	313.6	Her
220	16 42 22.422	+32 57 42.58	14.796	14.946	5.863	88.9	Her
221	16 45 45.216	+34 15 32.53	14.809	14.92	4.19	220.9	Her
222	17 01 49.512	+18 40 25.17	14.669	14.844	4.11	206	Her
223	17 05 21.561	+29 08 11.09	16.33	16.558	3.904	91.6	Her
224	17 06 00.275	+22 38 19.42	15.365	15.588	3.324	185.5	Her
225	17 08 28.215	+25 29 37.82	16.686	16.743	4.313	28.1	Her

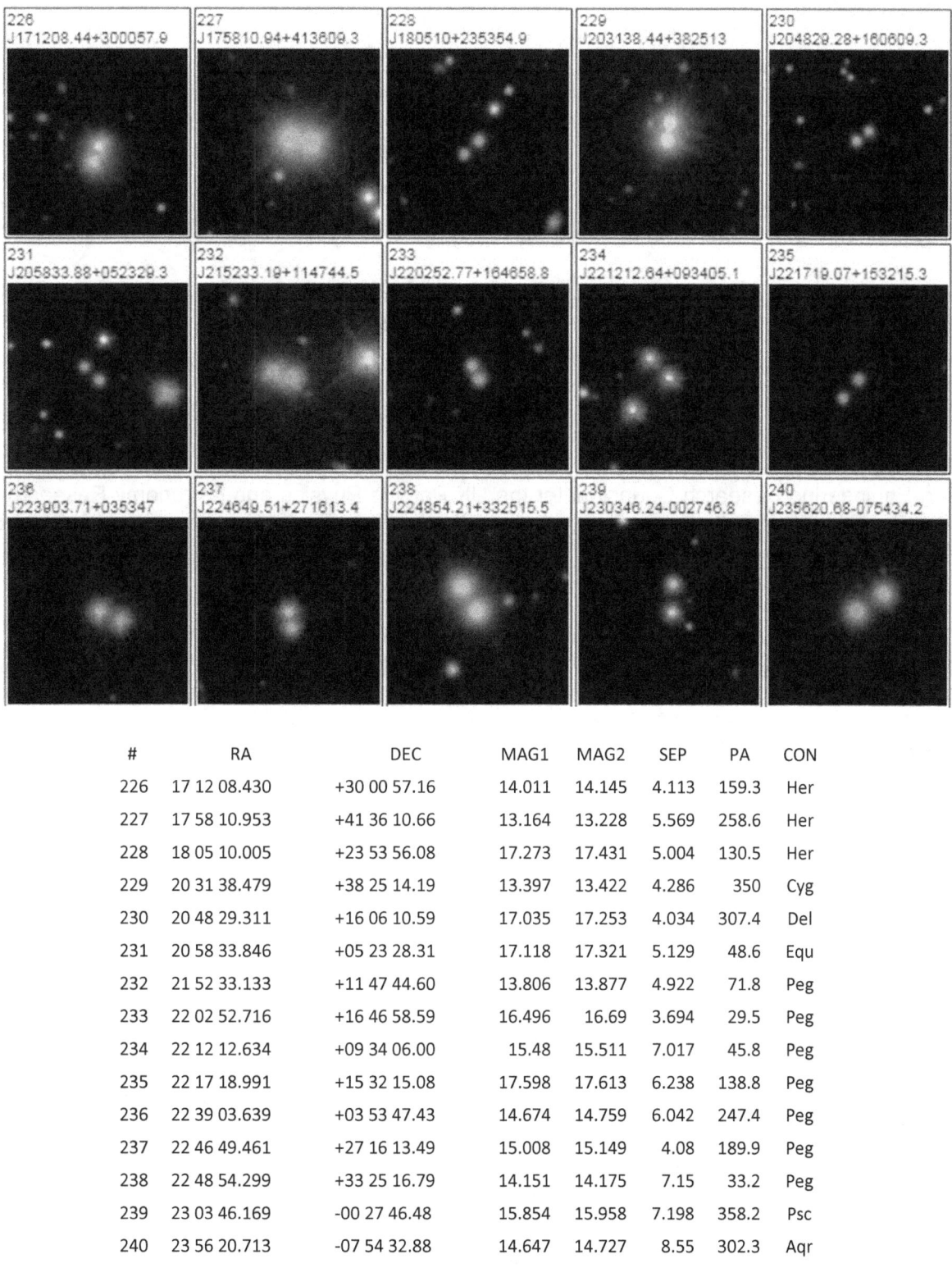

#	RA	DEC	MAG1	MAG2	SEP	PA	CON
226	17 12 08.430	+30 00 57.16	14.011	14.145	4.113	159.3	Her
227	17 58 10.953	+41 36 10.66	13.164	13.228	5.569	258.6	Her
228	18 05 10.005	+23 53 56.08	17.273	17.431	5.004	130.5	Her
229	20 31 38.479	+38 25 14.19	13.397	13.422	4.286	350	Cyg
230	20 48 29.311	+16 06 10.59	17.035	17.253	4.034	307.4	Del
231	20 58 33.846	+05 23 28.31	17.118	17.321	5.129	48.6	Equ
232	21 52 33.133	+11 47 44.60	13.806	13.877	4.922	71.8	Peg
233	22 02 52.716	+16 46 58.59	16.496	16.69	3.694	29.5	Peg
234	22 12 12.634	+09 34 06.00	15.48	15.511	7.017	45.8	Peg
235	22 17 18.991	+15 32 15.08	17.598	17.613	6.238	138.8	Peg
236	22 39 03.639	+03 53 47.43	14.674	14.759	6.042	247.4	Peg
237	22 46 49.461	+27 16 13.49	15.008	15.149	4.08	189.9	Peg
238	22 48 54.299	+33 25 16.79	14.151	14.175	7.15	33.2	Peg
239	23 03 46.169	-00 27 46.48	15.854	15.958	7.198	358.2	Psc
240	23 56 20.713	-07 54 32.88	14.647	14.727	8.55	302.3	Aqr

Acknowledgments

U.S. Government grant NAG W-2166. The images of these surveys are based on photographic data obtained using the Oschin Schmidt Telescope on Palomar Mountain and the UK Schmidt Telescope. The plates were processed into the present compressed digital form with the permission of these institutions.

The National Geographic Society - Palomar Observatory Sky Atlas (POSS-I) was made by the California Institute of Technology with grants from the National Geographic Society.

The Second Palomar Observatory Sky Survey (POSS-II) was made by the California Institute of Technology with funds from the National Science Foundation, the National Geographic Society, the Sloan Foundation, the Samuel Oschin Foundation, and the Eastman Kodak Corporation.

The Oschin Schmidt Telescope is operated by the California Institute of Technology and Palomar Observatory.

The UK Schmidt Telescope was operated by the Royal Observatory Edinburgh, with funding from the UK Science and Engineering Research Council (later the UK Particle Physics and Astronomy Research Council), until 1988 June, and thereafter by the Anglo-Australian Observatory. The blue plates of the southern Sky Atlas and its Equatorial Extension (together known as the SERC-J), as well as the Equatorial Red (ER), and the Second Epoch [red] Survey (SES) were all taken with the UK Schmidt.

Funding for the Sloan Digital Sky Survey IV has been provided by the Alfred P. Sloan Foundation, the U.S. Department of Energy Office of Science, and the Participating Institutions. SDSS-IV acknowledges support and resources from the Center for High-Performance Computing at the University of Utah.

The SDSS web site is www.sdss.org.

BY THE SAME AUTHOR

All are available from Amazon.com and from Amazon.co.uk

1800 new double stars for amateur observers

3600 celestial asterisms for amateur astronomers

Discover your own double star

Discover your own variable star

Identifying Common Proper Motion Binary Star Systems

Identifying Identical Twin Star Systems from the SDSS Data Release 10

www.ingramcontent.com/pod-product-compliance
Lightning Source LLC
Chambersburg PA
CBHW081414170526
45166CB00010B/3343